COMMENTS...

"This timely and valuable contribution led by Dr Isabelle Druc, a renowned ceramic specialist, brings the spotlight back to the study of pottery and its myriad relationships with people. This handy guide will be useful for both students and professionals interested in learning to investigate, with precision, the composition of raw materials and their transformation by people. It enables the identification and description of their choices that inform about the critical stuff (techniques, identity, values, landscape) of ancient cultures. "

—*George Lau*
Sainsbury Research Unit, University of East Anglia, UK

"This practical guide to macroscopic ceramic analysis will be of great value to many archaeologists, regardless of their knowledge of geology or previous experience with compositional analyses. The detailed descriptions and illustrations should be particularly helpful in the initial stages of analysis by facilitating the definition of paste groups (based on their constituent minerals, rock fragments, and textures); inferences about production techniques; and the selection of meaningful samples for more detailed petrographic and chemical studies. The book also serves as a clear, concise guide to terminology commonly used by ceramicists and geologists. "

—*Jason Sherman*
Dept of Anthropology, University of Wisconsin-Milwaukee, USA

"A milestone manual for the analysis of ceramic technology in the field and in the laboratory. This book, thanks to an excellent corpus of illustrations, provides useful tools for quantitative and qualitative identification of macro-components in the ceramic paste. Doing so, it supports scholars to better focus the analysis and selection of the samples from the very first stages of the research, and it also helps to direct decisions and field procedures. Finally in this study, organized as an atlas, the author makes an admirable effort to standardize nomenclature and classification terms, effort which will influence the work of many in the coming years."

—*Carolina Orsini*
Curator, Museo delle Culture, Milano, Italy

"This book is a detailed, thoroughly researched contribution to the visual analysis of ceramic paste and will be valued as a practical reference for archaeologists interested in ceramic production. Druc has produced an accessible, useful guide to identifying the primary materials selected by potters and the transformations these materials undergo in the firing process. The resulting work is an important complement to the existing archaeological literature on ceramic style and chemical composition and will help to advance studies of pottery manufacture and variability."

—Sarah Clayton
Dept of Anthropology, University of Wisconsin-Madison, USA

"Isabelle Druc and her technical collaborators Bruce Velde and Lisenia Chavez perfectly explain the somewhat abstract world of pastes and temper of ceramics seen through a portable digital microscope. The manual is an excellent methodological support for all archaeologists dealing with analysis of ceramic classification and technology, be it prehistoric or present pottery. "

—Olga U. Gabelmann
University of Bonn, Germany

"An excellent addition to the study of archaeological and ethnographic ceramics. Druc and her technical collaborators were able to synthesize both experience and analytical rigor in a must-have reference work for all interested in understanding human behavior from the study of ceramic pastes."

—María Beatriz Cremonte
CONICET-Universidad Nacional de Jujuy, Argentina

"Indispensable manual for all archaeologists interested in giving a thorough and objective basis to the discussion on the organization of ceramic production and distribution using the concept of *chaîne opératoire*. An important contribution to help the selection of samples for more advanced archaeometric analysis. "

—Krzysztof Makowski
Pontificia Universidad Católica del Perú

Portable Digital Microscope

ATLAS OF CERAMIC PASTES

COMPONENTS, TEXTURE AND TECHNOLOGY

Isabelle C. Druc

with the technical collaboration of

Bruce Velde

and Lisenia Chavez

Deep University Press

Blue Mounds, Wisconsin

Deep University Online !

For updates and more resources

Visit the Deep University Website:

www.deepuniversity.net

www.deepuniversitypress.org

Certificate in Deep Education:

www.deepuniversity.net/graduatecourses.html

For permissions, contact: publisher@deepuniversity.net

ISBN   978-1-939755-21-6 (Hardback)
        978-1-939755-07-0 (Paperback)

Library of Congress Cataloging-in-Publication Data

Keywords: 1. Archaeology. 2. Ceramic analysis. 3. Applied Sciences handbook. 4. Ceramic production. 5. Laboratory manual. 6. Druc, Isabelle C.

Target audience: students, professors and researchers in archaeology, ethnography, ceramic, geology and ceramic analysis.

Version 2

Front cover: In memoriam, Potter Manuel Hernández Suarez (?-2012), Cerro Blanco, San Pablo, Cajamarca, Peru. Background: paste of a cooking pot from Calpoc, Peru, made with coarse clay without addition of tempering material. 150x.

Back cover: Wares ready for sale, production of the family Ocas Heras and Felicita Aquino Minchan, Mollepampa, Cajamarca, Peru. Background: paste of a cooking pot tempered with gypsum 150x. Sorkun, Turkey.

How to cite this book

Druc I. 2015. *Portable Digital Microscope. Atlas of ceramic pastes. Components, texture and technology* (with the technical collaboration of B. Velde and L. Chavez). Deep University Press, WI.

ACKNOWLEDGEMENTS

I wish to thank the archaeologists, potters and institutions that authorized me to analyze and photograph the ceramics and the pastes illustrated in this manual. In particular, I am grateful to Yoshio Onuki Emeritus Professor of the Tokyo University, and Kinya Inokuchi University of Saitama and director of the Kuntur Wasi Archaeological Project 2012-2014; Jim Stoltman and Sissel Schroeder University of Wisconsin-Madison; Richard Burger, Frank Hole and Yukiko Tonoike Yale University; Veronica Acevedo University of Buenos Aires; Carlos Elera Arévalo director of the Sican National Museum; archaeologists José Pinilla Blenke and Wilder Leon. My thanks also go to ceramist Andrée Valley for sharing her expertise on firing, ceramic petrographer Beatriz Cremonte for her constructive remarks and Janine Kam for her help with text revision. The Culture Ministry of Peru is to be acknowledged for the permissions granted to analyze several of the ceramics presented here as thin sections' microphotographs.

Bruce Velde has had the kindness to review the English version of this Atlas, offering important advice based on his expertise in geology, while Lisenia Chavez collaborated in mineral identification in Chapters 3 and 4. I thank them both for their help.

NOTE

All the photographs in this manual were taken by the author except where specified. The paste photographs were taken with a digital Dino-Lite microscope. Microphotographs of thin sections were taken with a Nikon camera attached to an Olympus petrographic microscope.

The term digital microscope is used here to refer to a small portable microscope, without ocular lens, connected to the computer by way of a USB cable, which allows seeing the image on the computer screen. This is a digital modern version of the low magnification optical microscope.

CONTENT

1. INTRODUCTION

Describing the different characteristics of pottery fragments and entire vessels is a major part of most archaeological ceramic studies and an initial step to any further analysis. Surface treatment, color, and decoration are readily visible as opposed to the paste components. Very often a paste is defined as sandy, fine, coarse, brown, or with other general attributes. Several decades ago, Anna Shepard (1964) and Frederick Matson (1970 [1963]) recommended the use of a binocular microscope to study ceramics to help with typology and paste descriptions. However, this practice was never generalized or fully adopted among archaeologists. Now, with the digitalization and reduced size of hand-held microscopes equipped with a camera, image analysis program, and a USB connection to any computer, their use is becoming more frequent. It is easy to bring a digital microscope to the field, allowing for a much more practical and systematic analysis of the paste of the hundreds of ceramic fragments found on a site.

The macroscopic study of the paste helps examine a great quantity of sherds, establish typologies, group ceramics according to their main mineral components and texture, and select samples for further petrographic and chemical analysis, which reduces the cost of such analyses. Note that a macroscopic study, even with a digital microscope with good magnification, cannot replace the knowledge gained from a petrographic study. In some instances, very fine or low-occurrence minerals will be determinant in identifying provenance or in distinguishing between ceramic groups. Quantitative analysis is also best made with a thin section, and, as Anna Shepard stresses, one should not rely on a single photograph to quantify components in a paste (Shepard 1964: 519; see alternative options in Chapter 5). However, macroscopic analysis allows for a good estimate of the number of paste types in a collection, of the manufacturing techniques and possible raw materials used by the potter.

This being said, the description of the minerals, lithic fragments and other components in a paste is still limited unless the person examining the ceramics has some basic knowledge of geology and some reference collection to help him or her identify what he or she sees. This manual was written with this limit in mind and the intent to ease the work of the archaeologist in the first steps of ceramic analysis. It is hoped that this manual will be a very useful tool to help analyze ceramics in the field and in the laboratory.

The terminology, methodology and description of the minerals and rock fragments found in a ceramic paste are based on the common practices in geology and archaeological ceramic analysis. While no previous knowledge of geology is required to group ceramics in function of similarities in form, size, texture and basic mineral composition, it is very useful to take a geology class or to read about mineral identification. When we know what to look for, our eyes recognize better familiar patterns. It is also of great help to study the local and regional geology of your area.

Built like a mineralogical atlas, this manual presents images of ceramic paste under different magnifications, taken with a hand-held digital microscope (figure 1). The descriptions help recognize one mineral from another or certain manufacturing techniques. Notions of granulometry, texture and ceramic technology are introduced, which are linked to ceramic production.

1. Digital microscope Dino-Lite AM413ZTAS (grey cigar-like device) used here for the analysis of ceramic cross-sections. The microscope is connected to the computer by way of a USB cable. The white rectangular stage is an additional accessory providing transmitted light for basic petrographic analysis, but it is not needed for macroscopic analysis. The microscope has several LED lights and works in reflected light as do regular optical microscopes.

Note that there are many different portable digital microscopes on the market, and that it is even possible to take decent paste pictures with cell phones. Choose your microscope according to your needs and budget, but be sure to get one allowing different magnifications (e.g. 20x-200x), with a short nozzle to get close to your sample, and one that yields high-resolution pictures.

A ceramic is the end result of a series of decisions and actions, including the selection and preparation of the raw materials, elaboration techniques, and firing, all of which alter the original state of the materials used. This must be taken into consideration when interpreting the analysis data. While the minerals and rock fragments illustrated here are common in many ceramic pastes, they represent only part of the variety of inclusions found in a paste. The variety and variability that exist depend upon the geology close to the mining and production places and upon the tradition and practices of the potters. Furthermore, the relation between the place of production and the mineral components expected in a paste in view of the local geology may not be straightforward as the potter's behavior and ceramic production can be influenced by tradition and a whole set of socio-political, cultural and economic variables (see for example Arnold 1985, 2005; Gosselain 1992, 2000, 2008; Kingery 1982). Also, the majority of the examples presented here are taken from the work of the author, which introduces a bias as to the type of minerals and lithoclasts illustrated. This manual is only a first step to help in the identification of the paste components and in ceramic analysis conducted in the field with the new technologies at hand.

Ceramic analysis is often conducted to answer questions of provenance, to know if a ceramic is local or not, to understand better its manufacture, the level of technology, or the organization of production. This manual will not address these issues. However, it offers tools to do so. As well, a good knowledge of the local and regional geology is the basis of much interpretation of the ceramic data, and against which provenance hypotheses can be tested. The information provided here, when possible, relates to the geological origin or geological environment from which derive the inclusions seen in the ceramic paste. This does not refer necessarily to the location of the place of production.

2. METHODOLOGY AND TERMINOLOGY

2.1 Methodology

Surface preparation

An example of analysis protocol for ceramic cross-sections is given in Chapter 5. It is however important to present here the methodology followed to conduct a macroscopic analysis with an optical or digital microscope. This type of analysis is always best performed on a fresh cross-section, on a surface without contamination or deposit. To this end, it is necessary to break a little piece of the ceramic fragment, with pliers for example. This reveals the true color of the paste and allows for the examination of the inclusions and the texture of the clay matrix.

2.1 Example of fresh break (upper part of the cross-section) and superficial deposit (lower part of the cross-section). Bowl SG4, Kuntur Wasi, Peru. 70x.

One drawback of this technique is that breaking a chip of the ceramic does not yield a flat, smooth surface to study and photograph. There is a depth of field that is difficult to circumvent to obtain good images for publication. The problem is less acute for fine paste ceramics. One can also use a diamond saw allowing for a precise cut, such as those used to cut thin sections, but the process can be dangerous for the ceramic fragment, which can break into pieces, as well as for one's hands. A saw-cut cross-section allows a good vision of the inclusions, but the paste texture is somewhat lost. As for getting rid of a surface deposit, cleaning and brushing the fragment with a toothbrush and water is usually not sufficient. Scraping the fragment to obtain a flat surface does not work well either.

As one of the analysis objectives is often to obtain information about production and identification of different paste groups, the photographs taken with the microscope are used as archives to illustrate each group, internal variability or important compositional or textural details. The analysis can and should be done with the microscope looking at different parts of the cross-section, and not only based on an image taken from it. In certain cases, however, time limits and work restrictions or difficulty of access to the ceramic corpus, may require other work strategies. Then, it is important to augment the photographic archive, taking photographs at different magnifications and points of the cross-section, as well as taking notes of what you see and do. These images will allow you to conduct some modal or granulometric analysis providing that you have enough material to do so (see Chapter 5). Also, according to the character of the ceramic, if it is made with a coarse paste, without homogeneous distribution of the grains, you will need to examine a larger area than for a fine paste ware with homogeneous grain distribution.

Analysis

The methodology used for petrographic analysis of ceramics can be applied to the macroanalysis of ceramic cross-sections. It should be adapted to the corpus and objectives of each particular study. The following paste elements can be observed up to a certain point: mineral composition, granulometry, grain angularity, distribution and proportion of the inclusions, paste color, texture of the clay matrix, and the form, size and quantity of the voids or pores in the paste. The analyst will group the ceramic pastes that present similar characteristics. It is a set of elements (composition, texture, granulometry, percentages) that define a paste group. Once a profile of a paste group is defined, attribution to a particular group is easier. With experience and knowledge of the analysis corpus, one can classify many ceramic fragments quickly.

As mentioned earlier, sometimes it is the rare presence of some grains, or the combination of a set of elements or minerals that distinguishes a paste or a group from another. These can best be identified in petrography; in this regard the book and edited volume by Patrick Quinn (2009, 2013), or articles by James Stoltman (1989, 1999), offer good examples of what can be achieved in ceramic petrography, as do the classic works of Anna Shepard (e.g. 1942, 1965). Even so, if all cannot be identified or observed with a digital or optical microscope, much information can be obtained. It is thus important to register and document the particularities (color, form, size, frequency, etc.) that come to your attention, and describe them as best as possible, without necessarily giving a name to a mineral or rock fragment. If necessary, proper identification can be checked by a geologist or experienced analyst. This is true even for minerals that are common or very frequent like quartz and feldspars. It is good to describe their general aspect, if they are altered or not, if they present fine inclusions of other minerals, if they are rounded or angular, or if there is a mix of sizes and grain angularity in the same paste or for the same mineral type. All this information can be used later for data interpretation and even provenance analysis.

Another point of importance to consider when grouping ceramics with similar characteristics is the possible, and very probable, internal variability within each paste group. This is normal and can be linked to human and/or geological factors. This internal variability, however, should be less than the variability between paste groups, like the postulate expressed for chemical analysis by Weigand *et al.* (1977: 24). The criteria that define a group are partly dependent upon the context and objectives of the study and level of classification one wants to achieve. The internal variability of a group can be interpreted differently. A community of potters may get their material from the same source but process the raw materials with slight variations from one person to the next, in terms of material preparation or paste recipe. Also, and very commonly so, sources of raw materials (clay, sand, pyroclastic flows, alluvial sediments, etc.) may present internal mineral and chemical variabilities both vertically and horizontally. According to the depth, height and place in the vein or sediment mined, the granulometry or abundance of certain minerals can change (on this regard see Arnold 1985; Rye 1981; Shepard 1968; Velde and Druc 1999). Minor variability in a group can indicate a certain level of standardization in material preparation or paste recipe, while more important variability may suggest the presence of several independent producers.

These questions of variability in a corpus and subsequent interpretation of the analysis data point to an important aspect of ceramic analysis. The different steps of an analysis all involve selection processes by the analyst which are not or cannot be fully objective, and which concord to yield a certain representation of reality. These selection processes affect the selection of the corpus of analysis, the way it is analyzed, the attributes that will be later quantified, and the interpretations we draw from the corpus analyzed. This does not diminish the value of an analysis, but it reduces up to a certain point its interpretative potential. The more we know about the practices of the potters, the ceramic product, the factors affecting production, the behavior of the materials under certain conditions (firing, weathering, deposition, etc.) the better we will be able to analyze and understand the data collected.

The microanalysis of ceramic pastes is in great part of qualitative character, notwithstanding the quantitative aspect of granulometric analysis and estimation of components percentages in the paste. However, analysis reliability can be achieved even in qualitative analysis when rigor and consistency are applied. Geologists develop the ability to estimate the percentage of the different minerals constituting a rock, for example. This can be used for ceramic analysis as well, evaluating the percentage of individual crystals in the paste comparatively to the amount of lithic fragments, the amount of felsic versus mafic (see definitions below) minerals, etc. These estimates can be checked and experience allows one to reach coherence and save time.

Scales

The study of the texture of a paste includes looking at the granulometry of the inclusions, their degree of angularity (or sphericity), grain distribution and proportion, and size, form and abundance of the pores. A common granulometric scale used for classifying grain sizes is that of Udden-Wentworth, developed for sedimentology studies (see Folk 1965: 25). It is used in United States in ceramic studies. It is a logarithmic scale, with class limits expressed in phi φ scale or in millimeters and microns. Another scale in use to classify particles is the international ISO scale (see Rice 1987: 38). These scales differ for the superior limit of the clay class (2 μm in the ISO scale versus 3.9 μm with Udden-Wentworth) and medium-sand size class (0.63 mm versus 0.5 mm) (see table 1). In this manual the ISO scale is used.

Table 1. Granulometric scales (ISO and Phi φ) with correspondence in mm/μm (micron)

	ISO	Udden-Wentworth	φ
Very coarse sand	1-2 mm	1-2 mm	0-(-1)
Coarse sand	0.63-1 mm	0.5-1 mm	1-0
Medium sand	0.2-0.63 mm	0.25-0.5 mm	2-1
Fine sand	0.125-0.2 mm	0.125-0.25 mm 125-250 μm	3-2
Very fine sand	0.063-0.125 mm	0.0625-0.125 mm 62.5-125 μm	4-3
Silt	2-63 μm	3.9-62.5 μm	8-4
Clay	< 2 μm	< 3.9 μm	14-8

Scales and visual comparison charts for measuring or estimating grain angularity (or sphericity), abundance and degrees of sorting were adapted from sedimentology to ceramic analysis to help with the classification and description of the inclusions in archaeological pastes (see Cheel 2005; Folk 1951, 1965; Matthew *et al.* 1997; Rice 1987: 348, 380; Strienstra 1986; Velde and Druc 1999: 190-201). Figure 2.2 presents a simple angularity scale used in this manual. Figures 2.3 and 2.4 offer visual comparison charts for estimating the abundance of inclusions in a ceramic paste and degrees of grain sorting. For a classification of pores see the review by Rice (1987); however, pores are better studied in thin sections.

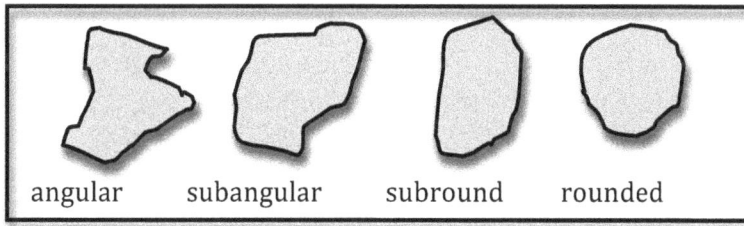

2.2 Angularity scale for grains, adapted from Müller 1964 in Strienstra 1986: figure 5.

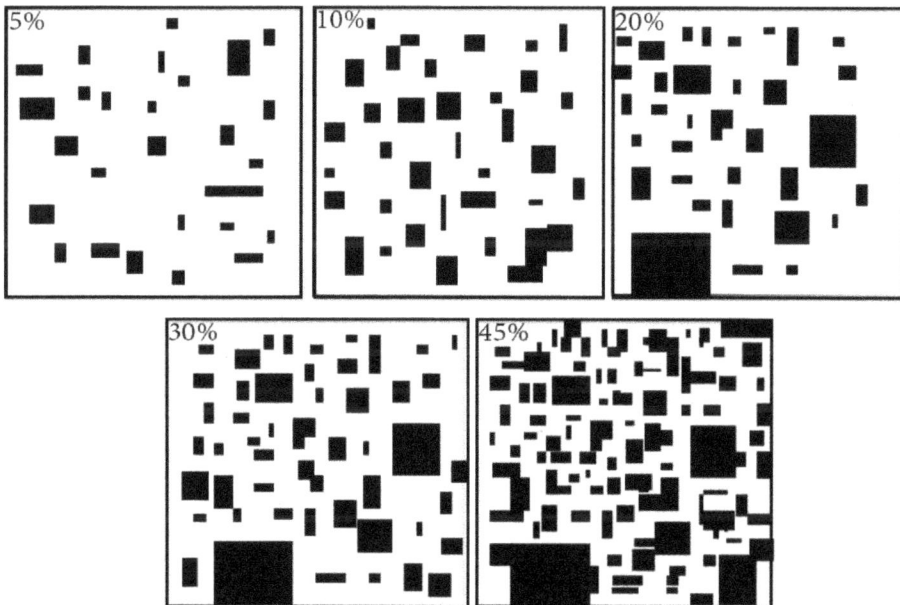

2.3 Visual comparison chart for grain abundance (redrawn and simplified from Folk 1951: Fig.1, percentages evaluated with the program JMicroVision, Roduit 2002-2008).

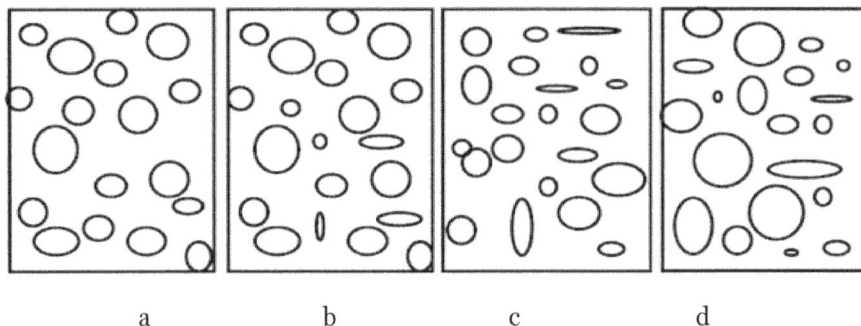

2.4 Visual comparison chart for degrees of grain sorting (after Anstey and Chase 1974 in Cheel 2005, figure 2.12), based on size and distribution of inclusions: a. very well sorted, b. well sorted, c. moderately sorted, d. poorly sorted.

2.2 Identification Problems

The photographs in this manual were taken with the common difficulties many analysts encounter. At times it is not possible or allowed to fragment a ceramic piece to obtain a fresh surface due to its archaeological or museum value. In that case, if nothing can be 'scratched' from the surface, the cross-section will probably bear some deposit that will mask the contour of the grains, or totally obstruct their examination. Even brushing the cross-section with a hard brush may not solve the problem. The information retrieved will be minimal but familiarity with the analysis corpus can somewhat palliate this problem and help in the identification of some of the paste components.

Granulometric analysis allows for objective comparison between ceramics or paste groups, and grain sizes are variables useful to study manufacturing practices and appreciate the work of the potter. However, a point to consider is the fact that a crystal or rock fragment is a tridimensional object. We may only see part of the grain according to how it is cut or embedded into the clay matrix, its position, or how much concretion covers its faces. Thus, when we measure an inclusion, the dimensions are rarely the exact length or width of the object. Even if we can usually classify a grain without much hesitation as fine, medium, or coarse, this introduces a range factor in the estimated grain size and possible erroneous classification in one of the granulometric classes. As well, there are several ways to measure grains: by their length, width, diameter, or area for example. Some minerals have forms that may dictate how they should be measured, micas are typically elongated, plagioclase often rectangular, pumices subround. Each analyst chooses a method; the important thing is to be consistent.

Also, the color of a paste may be affected by external factors that may bias its determination. Under certain light, and more importantly, at higher magnification, a paste can appear darker than it really is. If the analysis is consistent and conducted by the same person, this should not cause any problem. It is recommended to register the color of the paste when examining the cross-section at low magnification, or at least at the same magnification for all the fragments analyzed. One should keep in mind that the color of a paste is also affected by the raw materials used and the production process. The thickness of a ware, its porosity, paste granulometry, the presence of organic material, and firing atmosphere all influence the color of the paste. Matson (1970) and Rye (1981) give valuable information on this regard and how to interpret the range of colors observed in cross-section.

a) 85x

2.5 Difference in paste color due to change of magnification and light. The two photographs are from the same ceramic fragment taken at a) 85x, b) 145x. Bowl SG34b, Kuntur Wasi, Peru. If it were not for the coarse quartz crystal and some other components, one could think that these are images of different ceramics.

b) 145x

Finally, it is important to know that digital microscopes use an optical system that is distinct from those of optical or petrographic microscopes. This means that magnifications are not equivalent: a 40x with a petrographic microscope, for example, is not equal to a 40x with a digital microscope like the Dino-Lite.

2.3 Terminology

In ceramic analysis, the terms *inclusion* (be it mineral, organic or other) and *grain* (which usually is of mineral nature) refer to the non-plastic components in the paste or to the natural components in the raw material(s). It stands in opposition to the plastic components (e.g. clay particles) and clay-size minerals that compose the matrix or clay background of a paste. *Temper* is usually reserved to denote a material added by the potter to modulate the paste or clay base, such as sand, ash, chaff, etc.

Another term frequently used in this manual is *clast*, which means a fragment, and by extension lithoclast, that is, a rock fragment, or crystaloclast to designate a crystal fragment.

Felsic minerals in geology usually refer to light-colored minerals, such as silicates like quartz and the different feldspars (alkaline and plagioclase). Note that different types of feldspars exist according to their chemical composition. For convenience, here, the term *feldspar* is used to refer to the alkaline types (a-fd), which vary from potassium-rich (called K-fd, e.g. orthoclase) to sodium-rich members (Na-fd, e.g anorthose). *Plagioclases* are also feldspars but with calco-sodic compositions (e.g. albite, andesine, anorthite). The feldspar minerals display certain crystalline characteristics, such as shape, cleavage and twinning, which, in thin section and with a petrographic microscope, allow for their identification. In macroscopic analysis with a binocular or digital microscope this is more difficult. Thus it is better to use the general terms of plagioclase (for calco-sodic feldspars) or feldspar (for alkaline varieties).

Mafic minerals are dark-colored (e.g. green, brown, black), ferro-magnesian minerals, such as micas (e.g. biotite), amphiboles with hornblende being the most common, and pyroxenes. The latter are divided into clino- and ortho-pyroxenes, one frequent clinopyroxene being the augite (see Perkins 2002 or Winter 2010 for more information). It is advised to consult petrology books to look for the exact terminology to describe the different forms and textures of minerals and rock fragments.

In petrography, these different mineral varieties can be more easily distinguished than in a ceramic hand specimen. It is therefore often more prudent to use the terms felsic and mafic in the description of a cross-section unless a distinction can clearly be made. The amount of felsic and mafic inclusions, rock fragments and other components can be tabulated along with other criteria to group pastes according to similar composition or texture (see Chapter 5).

Intrusive (also called magmatic or plutonic) and ***volcanic*** rocks are ***igneous rocks***. Intrusive rocks are formed deep in the earth and crystals have time to grow and develop their characteristic faces. Volcanic rocks are extrusive, expelled during eruptions, and the crystals can be broken, small, of mixed sizes or as phenocrysts (large crystals in a much finer matrix). The rocks' origin and growth characteristics help distinguishing between intrusive and volcanic rocks. These are again subdivided according to their composition (mineral and chemical), such as the amount of quartz, plagioclase and mafic minerals present. Acid intrusive rocks have more silica (and thus felsic minerals, e.g. granite), basic intrusive rocks have more mafic minerals and few or no quartz (e.g. basalt, gabbro), and intermediate intrusive rocks have in-between compositions (e.g. granodiorite, andesite) (see geology books for details). Further description is also found in Chapter 3.

3. IDENTIFICATION OF COMMON MINERAL AND LITHIC COMPONENTS IN THE CERAMIC PASTE

3.1 Felsic Minerals

3.1.1 Quartz

3.1.2 Feldspars

3.2 Mafic Minerals

3.2.1 Micas

3.2.2 Amphiboles

3.2.3 Pyroxenes

3.3 Oxides and Hydroxides

3.4 Intrusive Rock Fragments

3.5 Volcanic Rock Fragments

3.6 Alteration of Igneous Rocks

3.7 Sedimentary and Metamorphic Rock Fragments

Chapter 2 presented the terminology used for the description of minerals and rocks. For more details about mineral characteristics and rock identification see mineralogy or petrology books (e.g. Perkins 2002; Winter 2010). The classification diagrams of rocks in geology manuals help identify a rock according to its mineral composition, texture, genesis and the granulometry of the crystals. However, a ceramic paste may include minerals and rock fragments of different geological origin and genesis, somewhat like a sediment of heterogeneous composition. As well, the fragmentation and alteration of the grains in a ceramic paste, due to the production process or the firing of the pot, may hinder proper mineral identification. Sedimentology manuals or petrology books on sedimentary rocks, such as the still current Folk 1965, yield very interesting information to help understand the relationship between the aspect or morphology of a grain or rock fragment and its origin. This, in turn, points to the possible resource areas used by ancient potters.

3.1 Felsic Minerals

To distinguish quartz from feldspar in an image taken by a digital or optical microscope one needs to look at the physical characteristics of the crystals. Quartz crystals display a vitreous shine, that can be greasy or translucent, and conchoïdal fractures. The crystals can have different forms, from angular with well-developed faces to amorphous. One main characteristic is that it is a very hard mineral that resists meteorization (i.e. chemical alteration by interaction with rain water). Feldspars, on the contrary, alter easily in particular upon contact with water, leading to hydrolysis and up to the formation of clay minerals for example. They present cleavage, twinning, and tabular crystal forms (or habits). They have hexagonal faces and some show scaled fractures, like the orthoses. The feldspars are a large group of minerals, varying in their chemical composition and mineral characteristics. As seen before, alkaline feldspars vary from a potassic (K) pole with minerals like orthoclase, to a sodic pole (Na), while the plagioclases vary from the sodic to a calcic pole with minerals such as albite (Na pole) to the anorthite (Ca pole). To distinguish those different types, one needs to do a petrographic analysis with a thin section or perform some chemical analysis. For hand-held ceramic fragments, it is better to only try distinguishing between alkaline feldspars and plagioclase.

3.1.1 Quartz

Quartz minerals can present themselves under different types. Here are a few illustrated.

3.1 Clear quartz, angular to sub-angular, of medium to very coarse sand-size, from a pyroclastic sediment used as temper in tradition-al ceramic produc-tion, Mangallpa Peru. MM15T. 150x.

3.2 Clear angular quartz. Reduced paste with short superficial oxidation layer. Quartz-rich, bimodal distribution, medium to coarse mineral and lithic inclusions in a fine-grained clay matrix. Large bowl CP11p, Kuntur Wasi, Peru. 165x (above) 80x (right).

3.3 Very coarse fragment of polycrystalline quartz (center) and fine biotite flakes (blue circle). The crystal measuring 2.072 mm presents a group of crystals of quartz without particular orientation. This distinguishes it, in part, from a sandstone fragment, which is a sedimentary rock, or a quartzite (metamorphic rock), which both show a different orientation and growth of their constituents. For example, in some cases, a quartzite may have suffered pressure or stress and the grains will be deformed, elongated or present some directional orientation. This is not the case with a polycrystalline quartz. Coarse olla, traditional production, Mina Clavero, Cordoba, Argentina.

3.4 Embayed quartz in volcanic paste, Tinaja SG53, Kuntur Wasi, Peru. 155x (right), 85x (below).

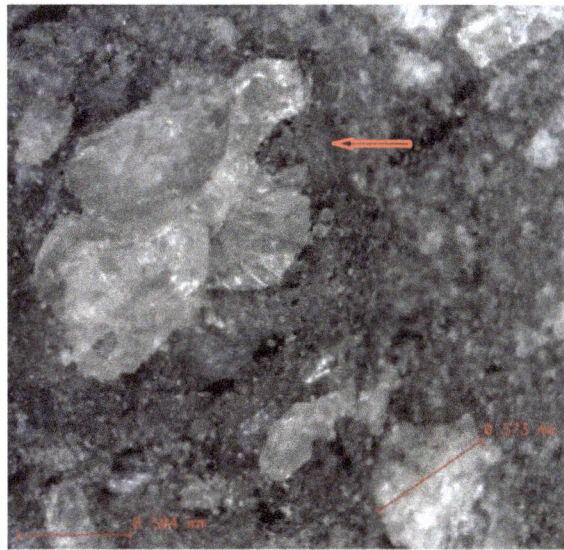

Embayed quartz (the red arrow points to one) is a characteristic of acid volcanic rocks. They do not have all their faces well developed. It is common to find them as individual crystals in ceramic paste made with sediments derived from these rocks and it is a good indicator of the volcanic origin of part of the sand or sediment used by the potter.

3.1.2 Feldspars

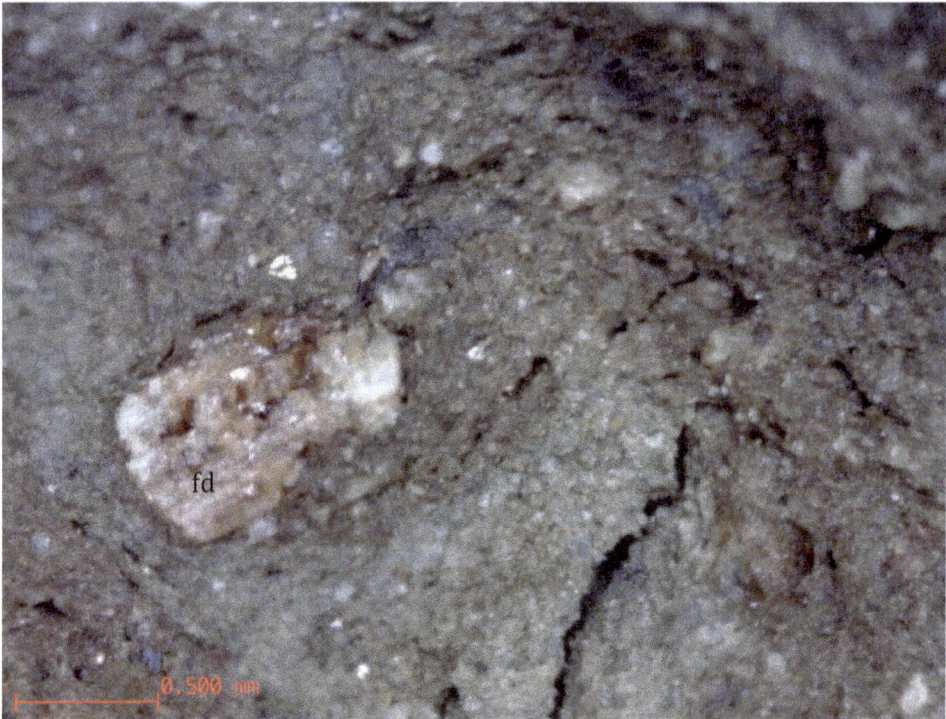

3.5 Coarse orthoclase crystal (0.967 mm long). Bottle base KW23, Kuntur Wasi, Peru. 150x.

In this photograph, the crystal presents a tabular habit, two well-developed faces and it is moderately altered to clay. The latter minerals are not present but have left holes filled with the oxides of a mafic mineral.

Note that the color of a crystal is not always a criteria to identify a mineral in hand specimen. Other physical properties are to be considered, such as chemical meteorization (alteration at the surface of the earth) for the feldspars.

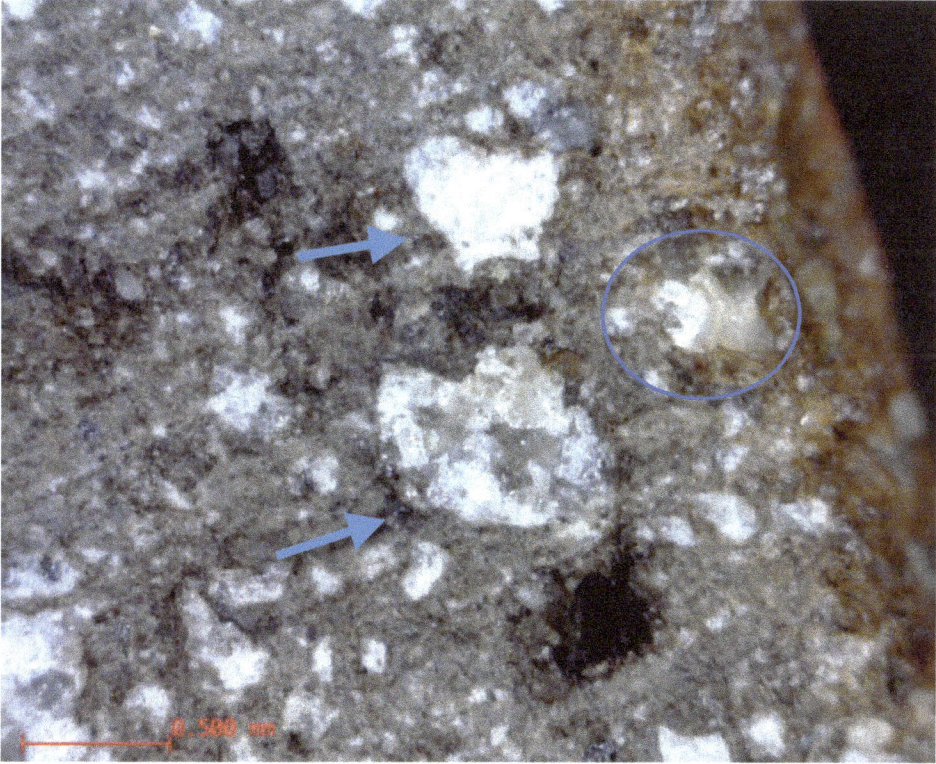

3.6 Feldspar. Large bowl KW26, Kuntur Wasi, Peru. 200x.

The fragment has been cut with a saw, which left parallel striation. The feldspar inside the blue circle presents characteristic scaled fractures. The coarse grains in the center are polycrystalline quartz (blue arrows).

3.7 Plagioclase (pl). Fine parallel lines are visible in the coarse plagioclase crystal illustrated here. These lines indicate the presence of twinning, characteristic of this type of mineral. Traditional cooking pot PR53, Cancharumi, Peru. 215x.

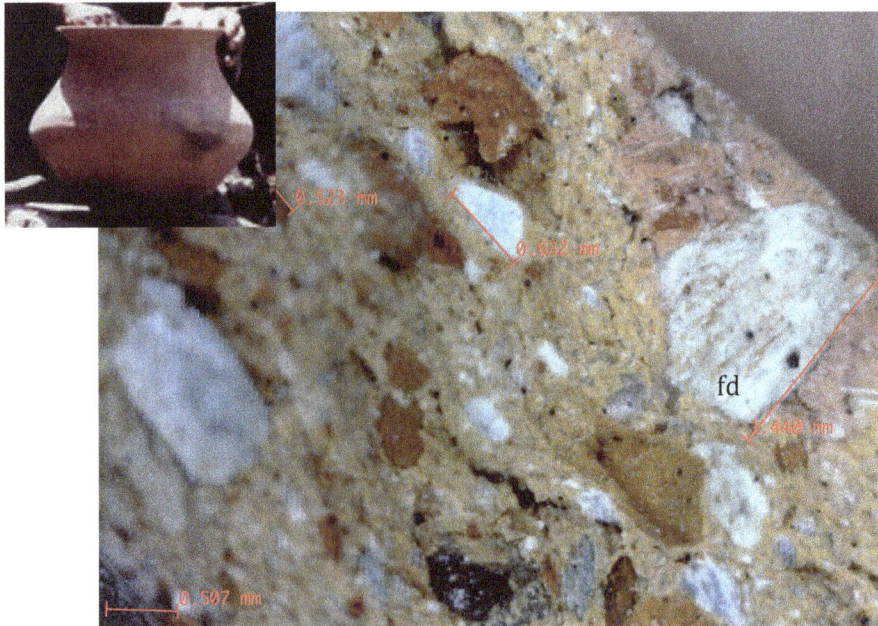

3.8 Feldspar (fd) altered to clay (coarse crystal to the right). Cooking pot PR41, Musho, Peru. 90x.

3.2 Mafic Minerals (iron- and magnesium-bearing)

Micas are sheet-like minerals, with cleavage (breaking plane - see glossary) in one direction. In a rock hand-specimen, for example, they can easily exfoliate, that is break off as thin plates. These are often shiny, a characteristic seen at the surface of many ceramic wares, in black micas (e.g. biotite) and white micas (e.g. muscovite) alike. Identifying characteristics of biotites are their brown to black color, a vitreous to silky shine, and well-developed faces, with a tabular shape or forming a hexagon if cut or broken perpendicular to the main, elongated axis. An oxide grain is also brown but distinguishes itself from a biotite by poorly defined faces and subround form.

Amphiboles are also mafic minerals but their habit is usually prismatic, with a light to dark green color, and a vitreous shine. Amphibole crystals can be elongated or present their basal face with their six characteristic sides forming angles of 120 degrees. It is common to see fragmented crystals, with only parts of their faces. Note that the angles of crystal faces are different characteristics than the inter-cleavage angles. The latter are best observed in petrography and are diagnostic features distinguishing a mineral family from another.

Pyroxenes, another category of mafic minerals, are prismatic as well but of darker color than the amphiboles, and shorter (stockier), with a mat shine. When they are well crystalized, they can have from four to eight sides. If the faces are quadrangular, the angle between them is of 90 degrees.

Mineral alterations affecting mafic and felsic minerals may impede correct identification, due to changes in the color and form of the crystals. In this case, they can be described as 'oxidized grains'. Their presence could be used as classification criteria to define a paste group. On the other hand, it is possible that human interaction, in particular the strategies used to 'age' a paste and the length of the process prior to forming the pots, may accelerate the weathering or oxidation of certain minerals in the mix. Firing the pots will also induce compositional and textural changes in many minerals according to the temperature reached. Finally, particular deposition contexts of the wares, or post-use, may contribute to the alteration of the mineral constituents of the paste. In principle, pre-Hispanic ceramics, in the Americas, as well as many Neolithic ones in Europe were not fired above 900 °C. Many modifications and the full collapse of the crystalline network, however, occur beyond that temperature. One notable exception is the case of calcite, which starts decomposing at about 500 °C. For more information, see Rice 1987; Rye 1981; Velde and Druc 1999.

3.2.1 Micas

3.9 Muscovite flake on the surface of a bowl. PR10, artisanal production, Cunca, Sechin Valley, Ancash, Peru. 150x.

Muscovites are colorless white micas, with yellow to brown tones, while phlogopites are gold colored. They stand out on the ware surface as they shine and reflect light. Biotites are common so-called black micas due to their dark green to black color, with a transparent to opaque shine. Muscovites are very common in terrigenous sediments and sedimentary rocks, while phlogopites are specific to metamorphic and ultramafic igneous rocks. More than the color, it is the geological environment in which they are found, as well as the presence of magnesium in the case of the phlogopite, which allows us to distinguish between those two types. Micas are natural inclusions in rocks and sediments, but they do not occur as 'loose' minerals. So one cannot say a paste is 'mica tempered', but rather that micaceous schist has been added to the clay base. Such schist can be added to the slip to give a glittery look (Beatriz Cremonte, personal communication, 3/14/14). This is seen in Sorkun, Turkey for example.

3.2.2 Amphiboles

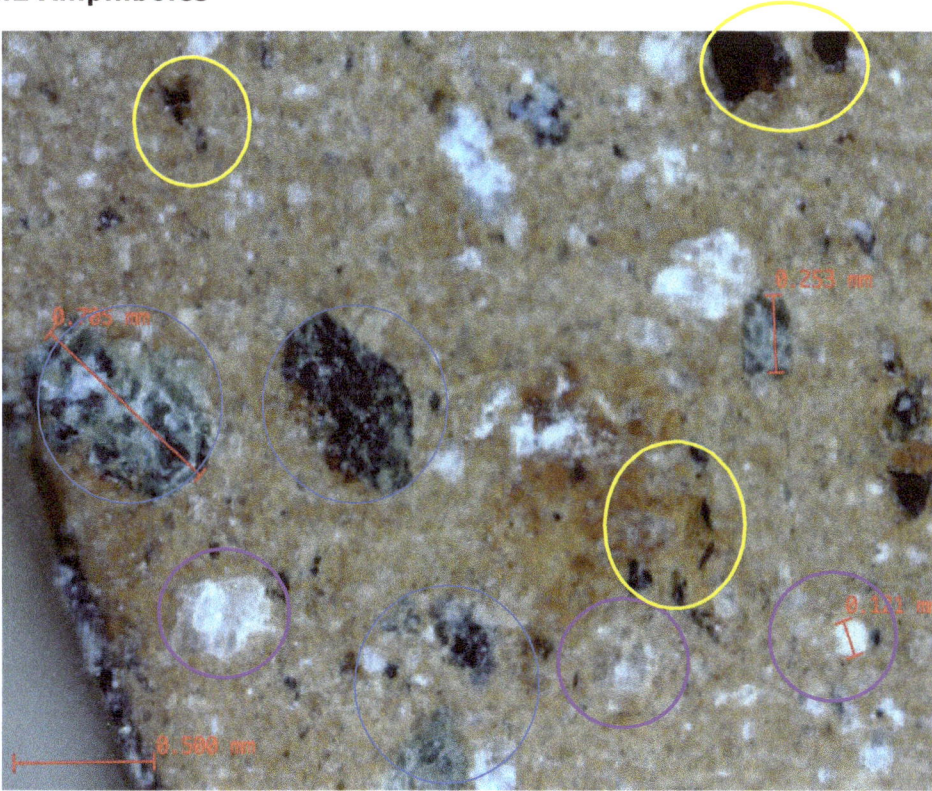

3.10 Crystals of amphibole (blue circles), biotite (yellow circles) and plagioclase (purple circles). Bowl, traditional production, San Marcos Acteopan, Mexico. 150x. Fragment cut with a saw for thin section. The dark green crystals with a vitreous shine (blue circles) slightly altered to chlorite are amphiboles. Their six faces are preserved forming 120° angles. The crystal in the center is probably a hornblende, a common amphibole. The darker crystals (yellow circles) are probably biotites based on their tabular form and color.

The white translucent crystals (purple circles) are plagioclases, per the vitreous shine and fine parallel stripes. In thin section under a petrographic microscope, these would be the polysynthetic twinning characteristic of the mineral. They maintain their faces and, here, are not altered.

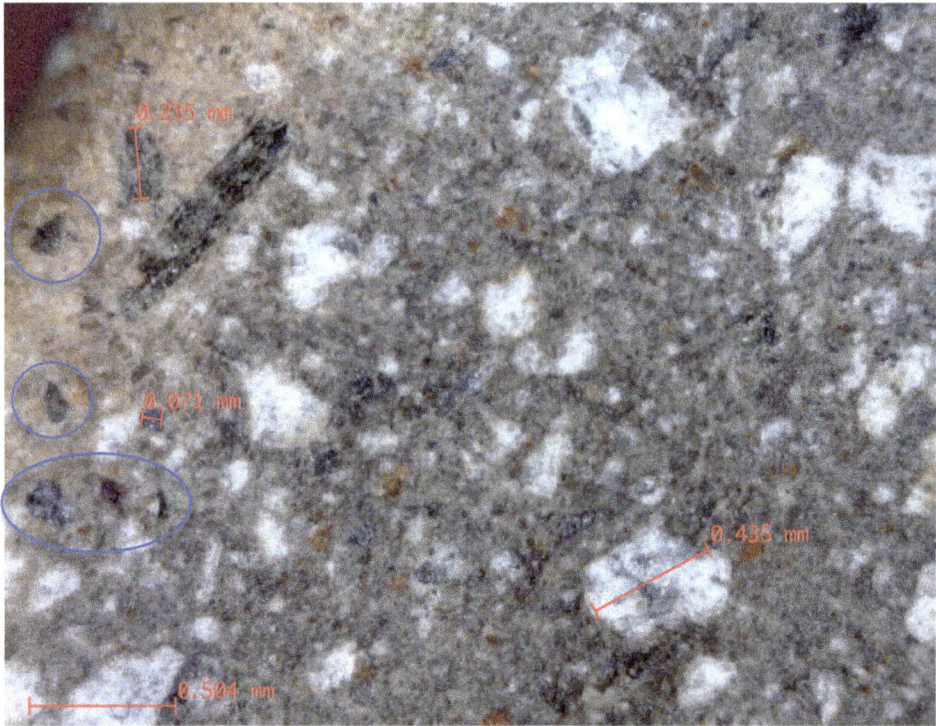

3.11 Hornblende. Decorated olla KW36, Kuntur Wasi, Peru. 150x.

The crystals with a tabular form are hornblendes. They present a prismatic form, are green and have a vitreous shine. The smaller crystals have been cut or broken and their basal face is visible (blue circles). In some, the 6-face hexagonal form is noticeable. The brown spots in the pictures could also be mafic minerals but with impregnation of oxides. At the right margin of the photo, one can see a feldspar crystal, still presenting two well-developed faces with a fine-scaled fracture.

3.12 Basal fragment of an amphibole (blue circle). The light green color of this crystal illustrates the tone variability that this mineral group can take, from green to black. The fine to coarse inclusions, reddish brown, are oxides and oxidized grains. A few quartz grains are also visible (e.g. the subround crystal measuring 0.583 mm) as well as feldspar and plagioclase crystals (yellow arrows). Bottle body CP66-2013, Kuntur Wasi, Peru. 85x.

3.13 Amphibole crystals (blue arrows). In the lower left part of the photograph, one can see an amphibole crystal (grain measuring 0.501 mm) with a basal cut showing the hexagonal form. Bowl KW25-2012, Kuntur Wasi, Peru. 70x.

In hand specimen, the difference between a mica and an amphibole can be seen in the form of the crystal faces, and more particularly the base, if cut or broken perpendicular to their long axis. If so, micas are usually hexagonal and very thin, like flakes, while amphiboles are more stocky prismatic crystals, thicker, and when eroded, they can take a subangular form.

3.2.3 Pyroxenes

3.14 Mafic minerals and oxidized grains (reddish grains with no definite crystal faces), in addition to quartz and feldspar crystals. The coarse crystal in the blue circle is possibly a pyroxene per its dull reflection and 90° angle formed by the side, which is different from the 120° seen in amphiboles. Bowl KW20p-2012, Kuntur Wasi, Peru. 160x (above), 80x (below).

3.3 Oxides and Hydroxides

3.15 Iron oxides (brown grains above) and hydroxide (black grain). Here the glossy black color and botryoidal shape of the center left grain distinguishes it as a hydroxide. Oxides and hydroxides are chemical compounds combining ions of oxygen (in oxides), and oxygen and hydrogen (in the case of hydroxides) with other elements to form salts and other compounds. Examples of iron oxides naturally occurring and often seen in ceramics are hematite (Fe_2O_3) and magnetite (Fe_3O_4). Examples of hydroxides are goethite ($FeO(OH)$, an iron III hydroxide) and gibbsite ($Al(OH)_3$, an aluminum hydroxide). Large bowl KW26, Kuntur Wasi, Peru. 200x. Saw cut fragment.

See figure 3.14 for an example of oxidized grain.

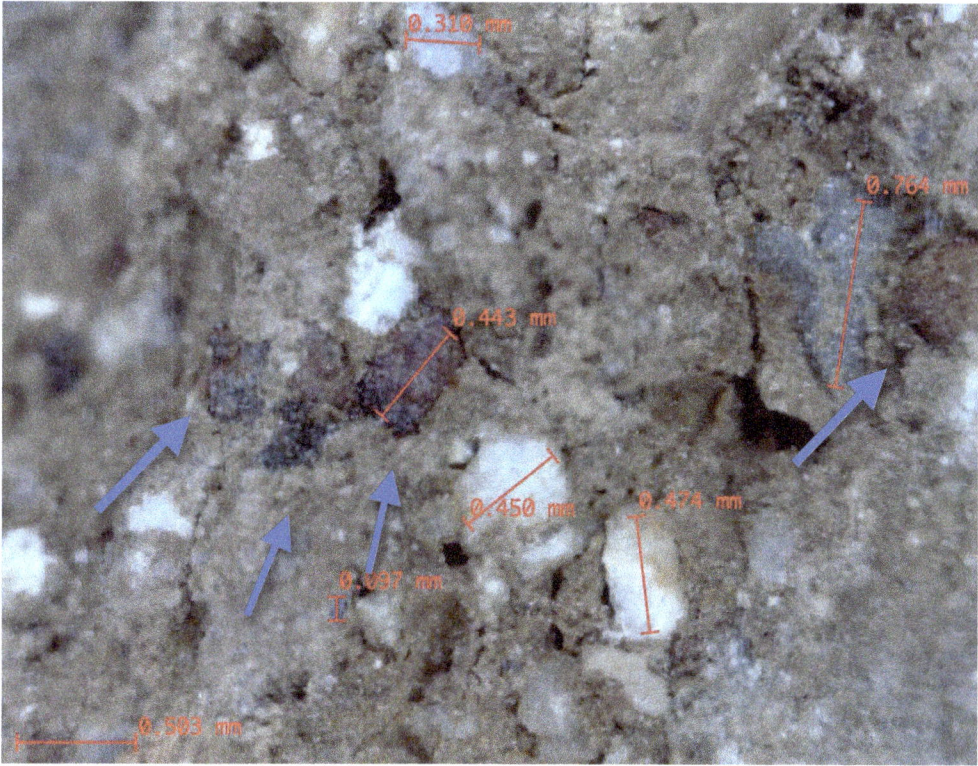

3.16 Mafic minerals that have suffered chemical meteorization (alteration and weathering at the earth surface), forming iron oxide impregnations. These appear as reddish to dark grains (blue arrows). We can also see quartz and feldspar crystals and a plagioclase crystal (just above the oxidized mineral measuring 0.443 mm). Bowl ID39-2012, Kuntur Wasi, Peru. 120x.

3.4 Intrusive Rock Fragments

3.17 Example of intrusive rock of intermediate composition (diorite), sampled near Chilete, at the Km. 10 on the road to Contumaza, Jequetepeque Valley, Peru, August 2012.

Intrusive rocks are classified according to the percentage of primary constituant minerals, such as quartz, potassic (K) feldspars, plagioclases, pyroxenes, amphiboles and micas. According to the composition of the minerals, and in particular the overall silicon content, a rock can be termed felsic or acid (with high amount of Si, like granites), intermediate (e.g. granodiorite and diorite), mafic or basic (low Si content and high magnesium – iron content, like gabbros) or even ultrabasic. The size of the grains and overall texture are also important criteria to classify a rock. Phaneritic would refer to a texture and grain size where the crystals in the rock are easy to distinguish without a microscope. How well are crystals crystallized is another criteria. Well-formed crystals, of fine to coarse sizes help differentiate intrusive (formed deep in the earth) from extrusive (volcanic) rocks. Refer to a geology book like Winter (2010) for details about rock classification, formation and identification or, for example, the Atlas of Igneous Rocks by MacKenzie *et al.* (1991) for descriptions and illustrations of different types of rocks in thin sections.

3.18 Coarse-size lithic grain and smaller fragment (red arrows) of felsic composition, probably derived from an acid intrusive rock. The smaller lithoclast left to the larger grain also has mafic minerals of different compositions (one green, two blacks). These lithoclasts can be described as holocrystalline (with more than 90% crystals), phaneritic (crystals visible to the naked eye), composed of coarse non-equigranular grains, of felsic composition (majority of quartz and feldspar), leucocrate (a color index for light colored rocks). These characteristics point to a rock of granitic composition. Bowl CP26-2012, Kuntur Wasi, Cajamarca, Peru. 170x (above) and 75x (left).

155x

3.19 Coarse-grained intermediate intrusive temper (with 52-60% of silica) of phaneritic texture, equigranular. The amount and kind of felsic vs. mafic minerals normally helps identify the type of rock in question. This can be best determined with thin section petrography. Bowl ID40-2012 with polychrome decoration Kuntur Wasi, Peru. 155x.

The oval reddish clast 0.355 mm long has the characteristic of an intrusive fragment highly oxidized. 170x.

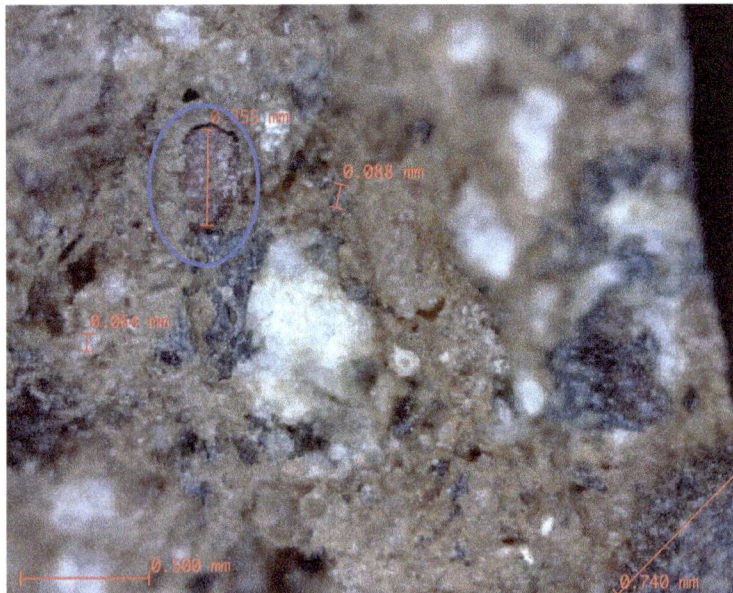

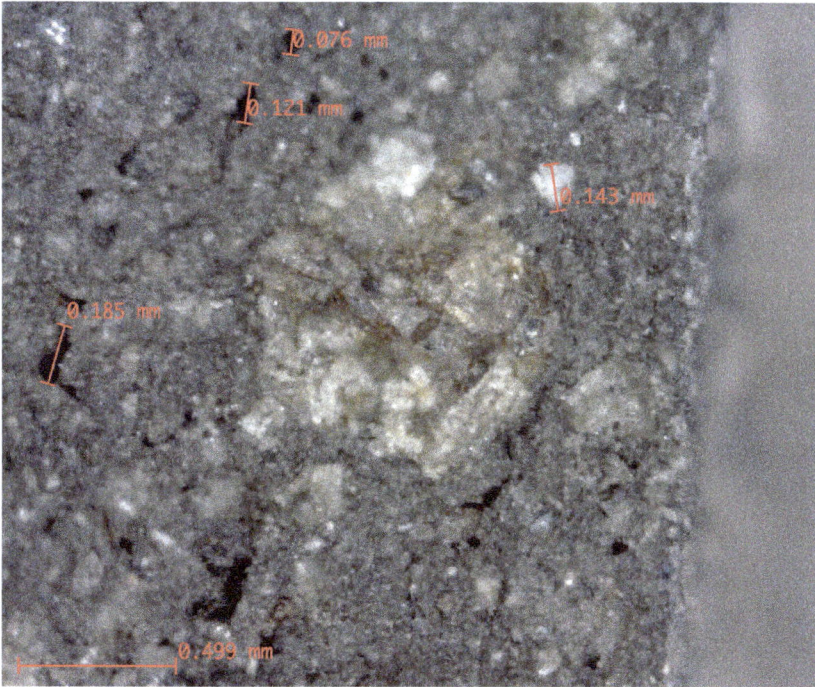

3.20 Very coarse subround lithoclast probably derived from an igneous rock. Oxidation does not allow a good identification. The amount of felsic and mafic minerals (circa 60:40) points to a plutonic fragment of intermediate composition (e.g. granodiorite). Smaller fragments of similar composition can be seen in the paste (right and below the large intrusive clast). Bottle CP41-2012, Kuntur Wasi, Peru. 175x (above), 80x (right).

3.21 Very coarse intrusive lithoclast (2.28 mm large) in which we can see crystals of feldspar, amphibole and pyroxene. The combination of felsic and mafic minerals suggests that this is a fragment from an intrusive rock of intermediate composition. These minerals, as well as quartz, are also found in the paste matrix. This *canchero* fragment CP56 from Kuntur Wasi Peru (see form illustrated below) is decorated with stamped circles, which might have been filled with red pigment. Some pigment is still visible on the left side of the fragment (red arrow). In comparison, the photomicrograph to the left shows the same ceramic fragment but prepared as thin section for petrographic analysis. The analysis with transmitted light and crossed polarizers at 40x allows one to see the mineral constituents of the lithoclasts in the paste, confirming the intermediate intrusive character of the lithoclasts.

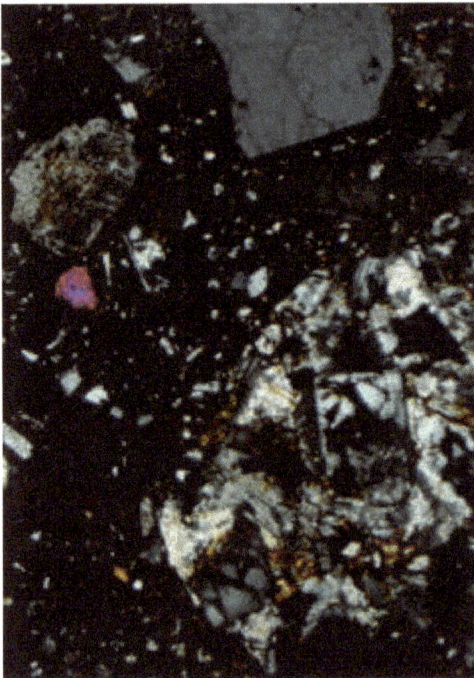

3.22 In this photograph, we can see several intrusive lithoclasts, subangular to subround, of fine to medium sand-size. The form and the variable size of the fragments suggest that they were part of a clastic sediment added to the clay base by the potter and not resulting from the process of mining and grinding fragments of the original rock. Bowl CP64, Kuntur Wasi, Peru. 150x.

Some sediments can accumulate fragments of crystals and rocks from the erosion of surrounding rock formations and soils. According to the type of rock, and the environmental and climatic conditions, the rate of alteration of these lithoclasts and crystals can vary. For example, volcanic rocks may become rounded and altered much more rapidly than intrusive rocks rich in quartz. Also, not all angular lithoclasts necessarily come from ground rock fragments.

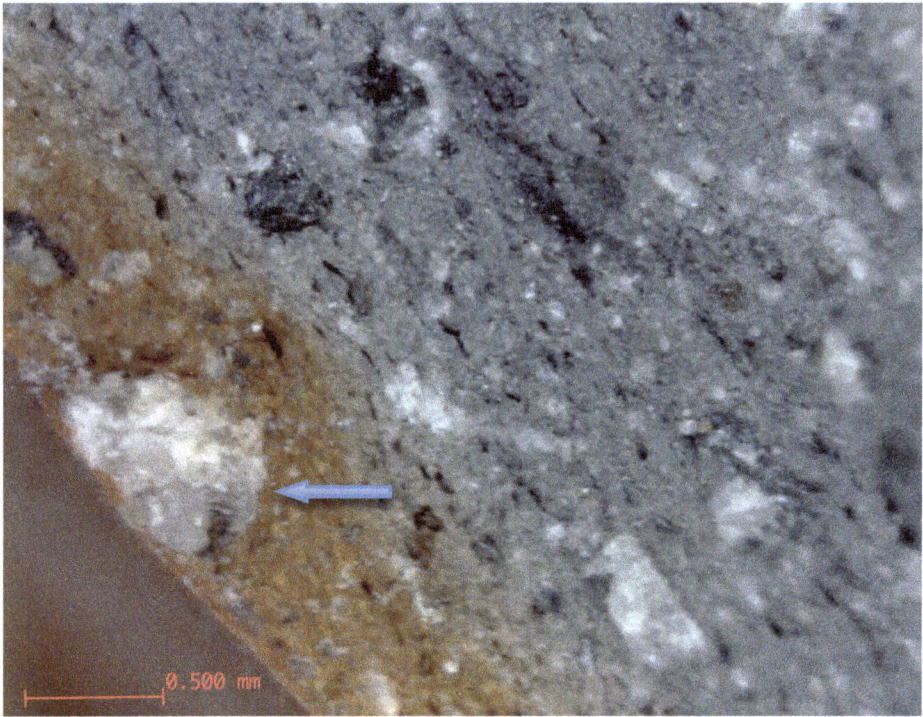

3.23 Coarse intermediate to acid (felsic) intrusive rock fragment composed of inequigranular coarse grains of quartz, feldspar, and mafic crystals. Bottle fragment PU156, Puemape, Peru. 150x. The photomicrograph below is from a petrographic thin section from the same ceramic, taken with crossed polarizers. The composition of the lithoclasts is clearer here and allows us to classify these as intermediate intrusive fragments of granodiorite composition. Transmitted light, 80x.

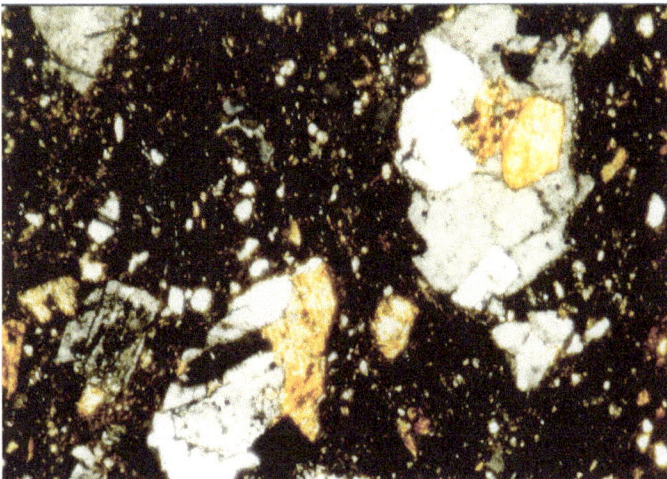

3.5 Volcanic Rock Fragments

Volcanic rock fragments are very common in ceramic pastes, in particular lava, pumice, and glass shards present in pyroclastic flows, which can fairly easily be mined. They may be found in deposits or sediments, and when not consolidated, the material crumbles and can be sifted without much prior pounding. The potter may also prefer these materials due to their resistance to thermal shock. Volcanic rocks are classified according to the size and composition of the rock fragments and crystals within the finer volcanic matrix. Lavas, pyroclasts, tuff (consolidated ash - but also a name given to pyroclasts) are categories distinguished by their texture and composition (see Winter 2010 or other geology books for a description of the different volcanic rocks). Simply put, these rocks have not had time to form large crystals with well-developed faces, unlike intrusive rocks, as they were more or less violently expelled to the surface of the earth. The crystalline texture of a volcanic rock may range from aphanitic (the crystals are too small to be distinguished with the naked eye in a hand specimen), to phaneritic (the crystals of the matrix are coarse enough to be distinguished) and porphyritic (big crystals called phenocrysts set in a matrix of much finer crystals). Volcanic rocks can be hypocrystalline (made up of crystals and glass) or present an important vitreous phase. In lava rocks due to the density of the flow, one can observe an alignment of the smaller crystals in the direction of the volcanic flow. Pumice is another type of volcanic rock used by traditional potters. Pumices are young rock fragments produced by a violent explosion. Their elongated structure (with angular extremities when not broken) and porous texture are characteristic. The texture can also be crystalline with a slight orientation of the crystals. Pumice fragments are clear or light colored due to the predominance of felsic minerals. Pyroclasts are formed of broken crystals, angular lithoclasts of different sizes (they have not been transported far) and can include glass shards and pumice. Pyroclastic deposits can be consolidated or not.

3.24 Examples of pyroclastic rocks collected close to Sangal, Cajamarca Dept, 2013.

3.25a Pumice fragment (coarse, subround clear clast, yellow arrow). Large jar (*tinaja*) SG53, Kuntur Wasi, Peru. 150x. One coarse quartz phenocryst with embayment can be seen within the pumice.

A pumice is a pyroclastic fragment of acid composition, composed of glass and with a vesicular texture. It presents usually an elongated form and can enclose crystals of quartz, plagioclase, biotite hornblende, or rock fragments. The image below illustrates a pyroclast with same composition as that seen in the ceramic paste above. It comes from a quarry used by potters to obtain tempering material (see Chapter 4, fig 4.2).

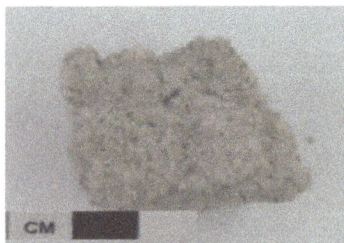

3.25b Pyroclast, Jancos Alto, Cajamarca, Peru.

3.26 Volcanic paste with various dark grey to black lava clasts (blue arrows). These lava fragments have an afiritic texture (without phenocrysts but with equigranular crystals). The fragments are slightly oxidized, have an important mafic component and a few plagioclase crystals. Another lava fragment (black lithoclast below) comes from a felsic pyroclast also with afiritic texture. These would typically contain plagioclase and feldspar of sanidine type. It is slightly oxidized. Bowl KW19-2012, 160x (above) and 80x (left). Kuntur Wasi, Peru.

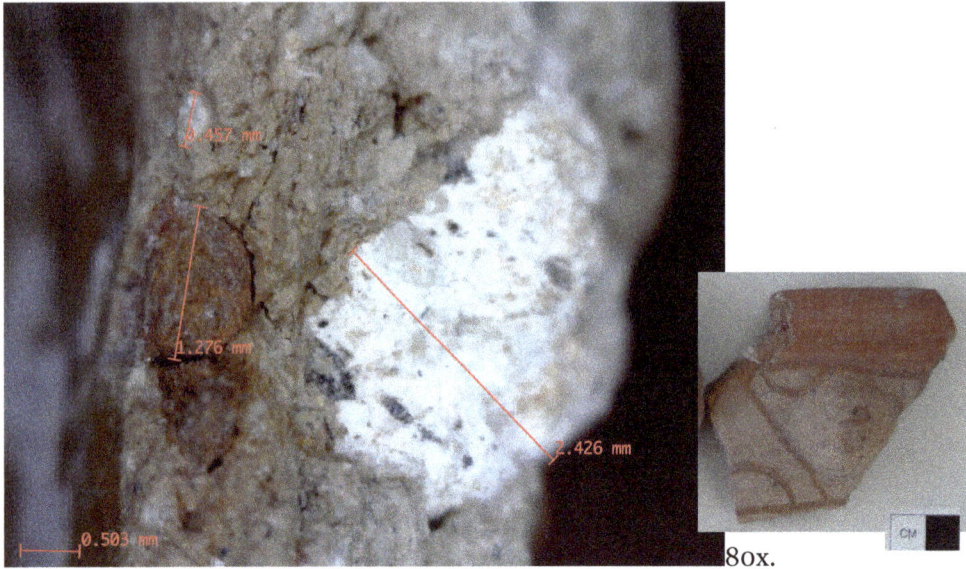

80x.

3.27 Coarse paste with very coarse lithic fragments. The white lithoclast is a pyroclast, with many mafic and felsic phenocrysts in a cryptocrystalline background (with microcrystals barely visible with the microscope). The dark greenish black elongated crystals within the pyroclast may be amphiboles. Bowl CP33-2012, Kuntur Wasi, Peru.

3.28 Pyroclasts with mafic phenocrysts. *Tinaja* CP17, Kuntur Wasi, Peru. 70x.

3.29 Lava fragment (round, blue arrow), volcanoclast (black clast), angular quartz and plagioclase crystals. *Tinaja* SG53, Kuntur Wasi, Peru. 155x.

3.30 Medium to coarse porphyritic volcanic fragments, angular to subangular (green circles) with phenocrysts of plagioclase (white striated subangular crystals) visible in the coarser volcanoclasts. Bottle KW37p-2012, Kuntur Wasi, Peru. 155x.

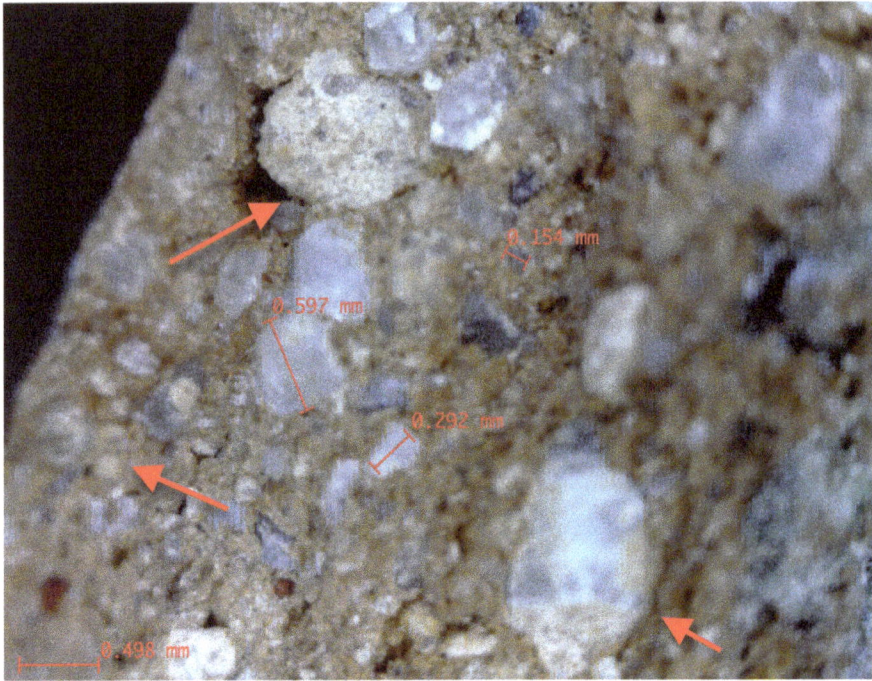

3.31 Fine to coarse round to subround pyroclasts (light-colored grains, red arrows) with abundant fine to coarse phenocrysts of quartz (clear white), plagioclase (milky white to grey white), and occasional biotite (fine brown flakes). Seen in thin section (right, plain polarized light, 40x), the pores of the pumice and the vesicles in the glass shards are flattened, parallel to the direction of the pyroclastic flow. This texture is caracteristic of pumice fragments. In a hand specimen, this is harder to see and it is better to identify the fragment as just a pyroclast. Olla MA24, Mangallpa, Peru, traditional production. 90x.

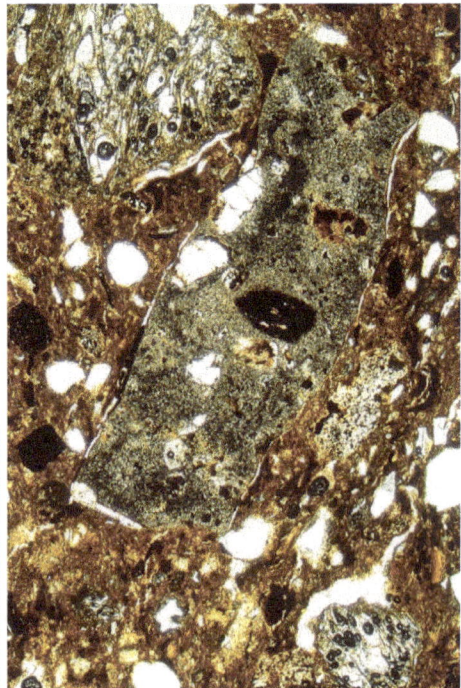

3.6 Alteration of Igneous Rocks

At the surface of the earth, rocks are subjected to chemical meteorization due to environmental chemical changes that modify their mineralogy, texture and chemistry. Different processes can happen, from oxidation to carbonatation, dissolution, hydration, lixiviation (leaching of soluble constituents by percolation) and hydrolysis. Igneous rocks that have been hydrothermally altered show an indication of the mineral most abundant, such as silicification (dominance of silica or quartz), argillization (dominance of clay minerals), etc. Ashes and glass in volcanic rocks are particularly affected by meteorization and alter rapidly to clay minerals, siliceous microcrystalline grains, or very fine feldspars (Cuadros *et al.* 2013; Folk 1965). Rocks that suffer alteration or destruction of their mineralogy, and major hydrolysis, can turn into a siliceous mass. We will see below two examples illustrating volcanic rocks that have been altered.

3.32 The coarse rounded grain in the center of the photograph is probably a pyroclast, silicified and presenting moderate oxidation. Olla CP34, Kuntur Wasi, Peru. 75x.

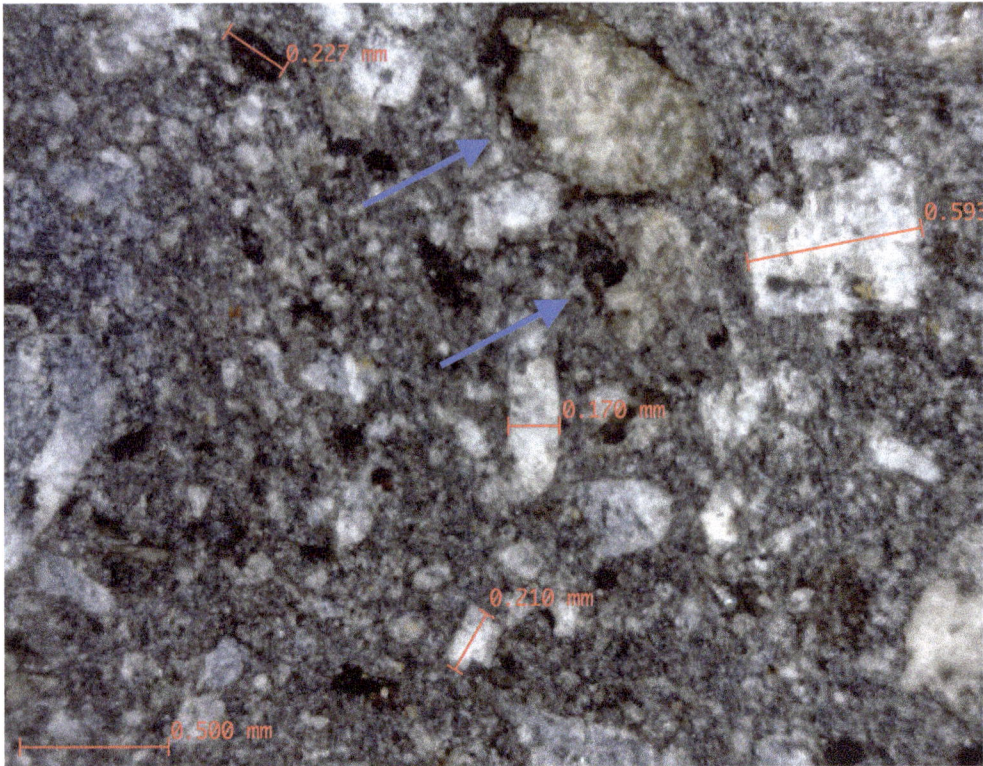

3.33 The subrounded grains (blue arrows), light-colored, with no clear definition of their constituents may be altered volcanic clasts or crystalized spherulites that come from acid volcanic rocks, such as rhyolites with a small nucleus of crystallization. The photograph also shows many feldspar crystals (e.g. coarse rectangular crystal close to the upper right margin). Olla CP59, Kuntur Wasi, Peru. 90x. Saw-cut fragment.

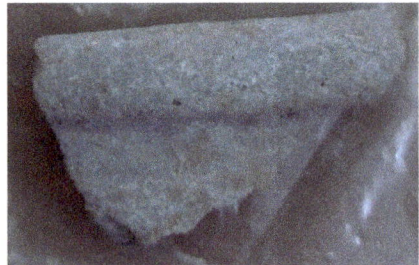

3.7 Sedimentary and Metamorphic Rock Fragments

Sandstones are one of the clastic sedimentary rocks that are often found in ceramic pastes, as they are common components of sands. They are composed in majority of sand-size grains (smaller than 2 mm) of quartz, feldspar, lithoclasts and/or organic material, in a matrix rich in silica or calcium carbonate. Sandstones with a high percentage of quartz are called quartz arenite (quartzite is the metamorphic equivalent, where the grains take all the space and not cement or matrix can be seen). Sandstones form from the action of weathering and alteration of preexisting rocks (igneous, sedimentary and/or metamorphic). They exist in great variety and color is not a diagnostic.

Metamorphic rocks are recognized by their texture, showing a recrystallization of the minerals in a solid state (blastic texture). A quartzite for example presents a granoblastic texture with a high component of quartz grains.

3.34 Sandstone fragments (sedimentary), grey angular clasts (blue arrows) of fine to medium sand-size. Neckless olla An66a, Ancón, Peru. 150x.

3.35 Very coarse fragment of quartz arenite (sedimentary rock) composed in majority of polycrystalline quartz and minor subround mafic minerals. Bottle An51-173, Ancón, Peru. 90x.

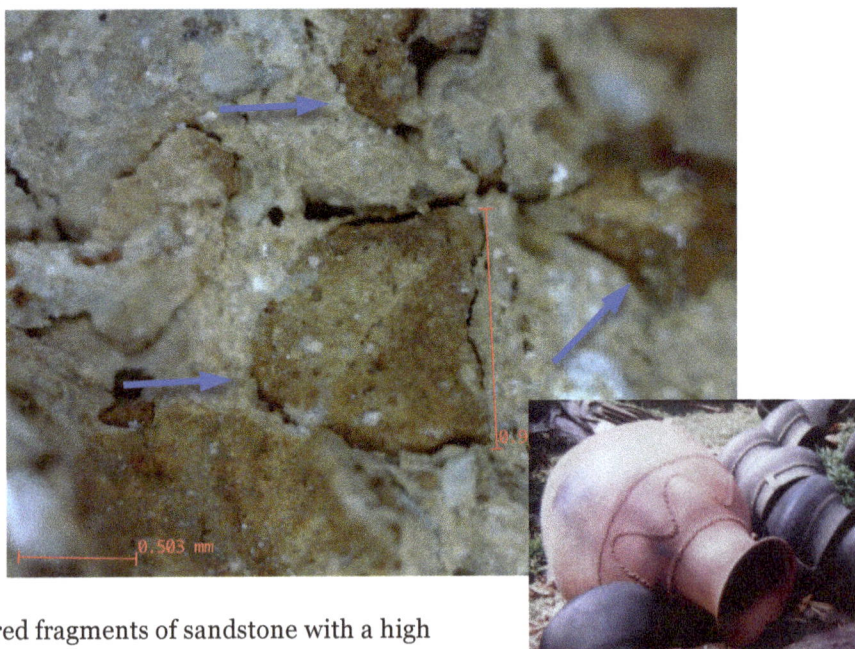

3.36 Altered fragments of sandstone with a high matrix component and few individual minerals. Traditional olla from Pariahuanca, Peru. 160x. The black colored wares (right) have not been fired yet. Once fired they will get the reddish tone of the large jar next to them.

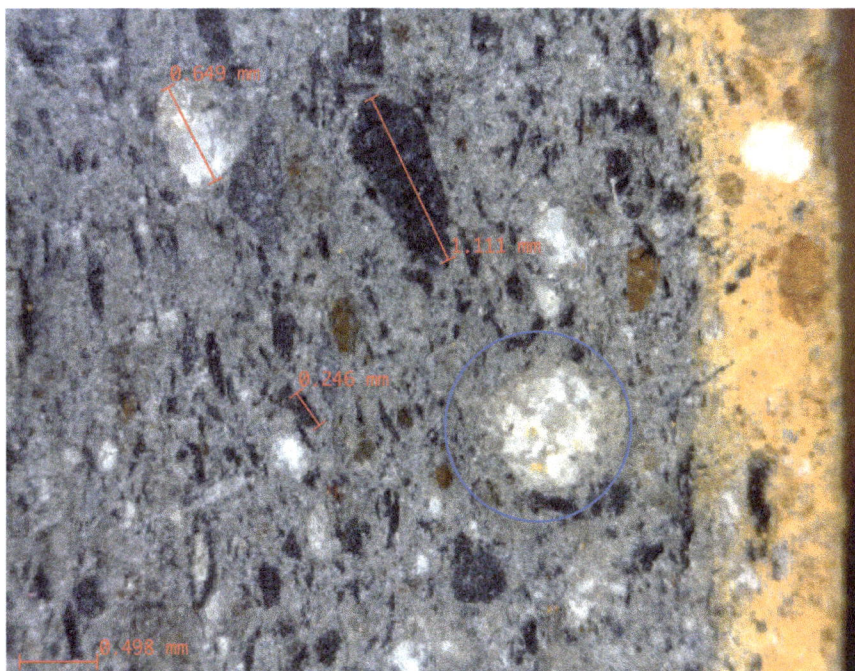

3.37 Slate-tempered ware. Fine to coarse sand-size crushed slate (metamorphic) fragments (black angular to subangular inclusions with a lamellar texture). Other minerals and lithoclasts present are: quartz, feldspar, oxides, acid intrusive clasts, quartzite and shale fragments. Reduced paste with oxidized surface. Cooking pot MAR10, saw-cut fragment for thin section, Marcajirca, Peru. 90x.

3.38 Same ware as in 3.37, but seen in thin section, in plain polarized light. The crushed slate fragments are black, nearly opaque. The coarse, left grain is a quartz sandstone. 40x.

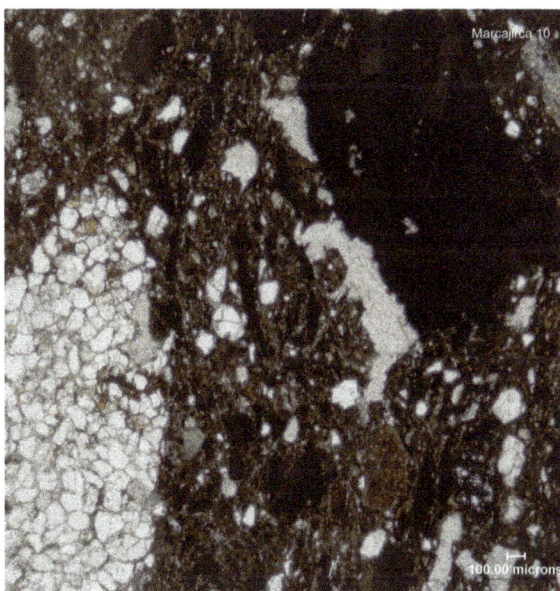

3.39 Slate-tempered ware. Fine to coarse sand-size black, grey and purple slate fragments (angular to subround inclusions with a lamellar texture). 30x. These metamorphic slate fragments are typical of the Puncoviscana Formation in the Quebrada de Humahuaca (Acevedo 2013). The white specks are salts from post depositional context. Bowl 31-68, Pukara de Tilcara, Humahuaca Quebrada, Jujuy, Argentina, *Negro sobre Rojo* style, pre-Incaic period. Ceramic fragment hosted in the collections of the Museo Eduardo Casanova, Tilcara, Jujuy, Argentina. Paste and ceramic photos: Veronica Acevedo, Universidad de Buenos Aires. Paste photograph taken with a Microlab digital microscope.

4. RAW MATERIALS AND CERAMIC TECHNOLOGY

The following pages illustrate various types of raw materials, pastes tempered with organic and mineral tempers, evidence of manufacture, examples of textures and some effects of the firing atmosphere as seen in a ceramic cross-section. These examples show the type of information obtained from macroscopic analysis of ceramics as well as raw materials. Usually, the clays used by traditional potters are not pure and may contain many fine to coarse non-plastic materials, organic and mineral in nature, which inform us on the type of sediment used. Note that the analysis of clay minerals cannot be done with a digital or petrographic microscope, as their resolution and magnification power are not high enough. Nevertheless, a preliminary study of the texture, composition of the non-plastic inclusions and their granulometry is informative of the type of clays used or available in a region. Sediments or clays suitable for ceramic manufacture should also be collected as comparative materials, keeping in mind that two or more raw materials, refined or not, can be used in a paste recipe. These comparative materials can be prepared, sieved with different mesh sizes, mixed using different ratios and fired at varying temperatures. Their study as test tiles with the digital microscope helps in the comparison with the archaeological ceramics.

The color of a clay body and ceramic surface depends upon several factors (see Shepard 1968: 16-22), amongst which are the percentage and state of the iron present in the paste, firing atmosphere and temperature. The composition, type of material and porosity of the paste are also important variables that combine with temperature and firing atmosphere to yield the product we see. These factors should be kept in mind when examining the surface and cross-section of a ceramic. Also, a bimodal grain distribution (e.g. presence of fine and coarse sand-size inclusions without the medium fraction) may indicate the use of two raw materials, or the addition of a coarser fraction to the clay base, including the re-insertion of part of the fraction that was initially separated from the original material by way of decantation or levigation. Such preparation is common in San Marcos Acteopan, Mexico (Druc 2000), allowing the potters to modulate the amount of coarser material they add in function of the ware to be produced. The use of one source material to obtain two or more base ingredients also insures their compatibility. Considering that the potters can modify their raw materials, estimating the percentage of inclusions per granulometric classes can be useful for characterization, comparative purposes and information about material processing. Similarly, paste texture and granulometry of the inclusions give essential information about the technological tradition of the ancient potters.

4.1 Raw Materials

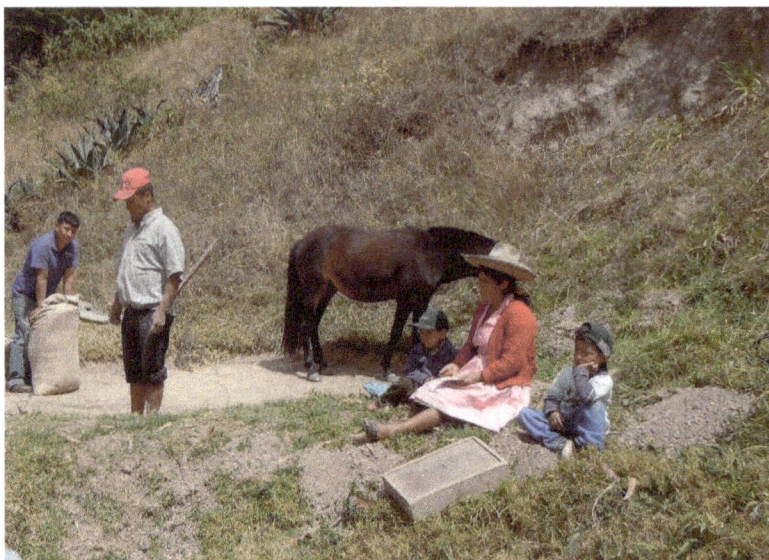

4.1 Clay mining, Mangallpa, Peru. Potter Santos Tanta Sánchez (with red cap), his family and a helper. August 2013.

4.2 Temper mining area, pyroclastic material, Mangallpa, Peru.

The above two materials (clay and temper) are mixed in proportion of 1 measure of clay for 2 of temper. Many ceramics from the archaeological site of Kuntur Wasi illustrated in this manual present the same pyroclastic composition.

4.3 Raw material used as temper in traditional production, Mangallpa, Peru. (a) Wall of consolidated volcanic pyroclastic deposit (same place as 4.2) and (b) close-up view with volcanic tuff fragments, pumice, quartz, plagioclase, and subvolcanic lithic fragments. In the ceramics made with this tempering material, the coarser fraction (superior to 1 mm) has been eliminated. 4.3 a and b: 90x.

a) 90x

b) 90x

4.4 Medium to coarse clay, Sangal, San Pablo, Peru: (a) raw clay with non-plastic, fine inclusions and a few medium to coarse-size quartz and sandstones; (b) unfired test tile, sieved clay (without the medium and coarser sand fractions) mixed with water. This clay is composed principally of illite clay minerals (Druc *et al.* 2013).

a) 90x

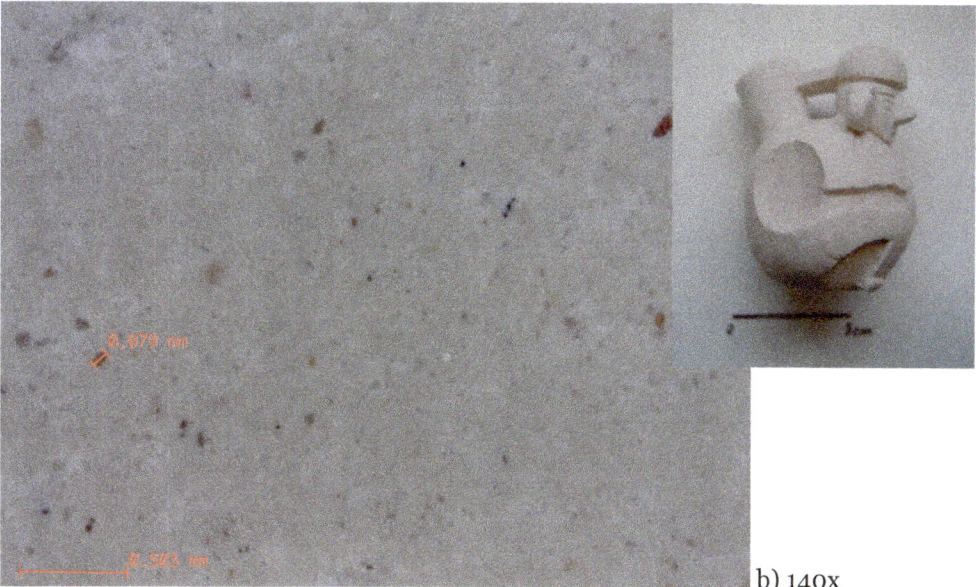

b) 140x

4.5 a) Raw kaolin, Callejón de Huaylas, Peru. b) Unfired kaolin paste without addition of temper. Imitation of a Recuay vase provided by archaeologist Wilder León, 1998.

Pure kaolin does not contain iron. The clay is white and does not change color upon firing.

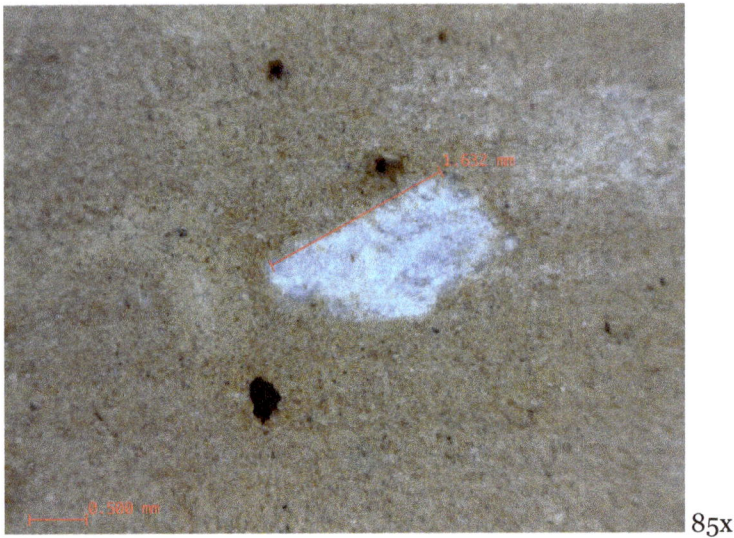

85x

4.6 Fired test tile, made of loess (silt and clay). No. AY2008 tN01 - ALN09A6. Loess collected near Linjiazhuang (ALN), archaeological site of Yinxu (Anyang), China. Very fine clay with few non-plastic inclusions, like the feldspar crystal above (white, very coarse grain) and few oxides (black inclusions).

110X

4.7 Paste made with loess without added temper. Guan type jar HB99-5, *Grayware*. Site of Huanbei, China. The change of color of the loess (grey) as compared to the light tan color in figure 4.6 results from the firing atmosphere. The coarse inclusions in the original raw material were eliminated.

85x

4.8 Mold fragment for bronze production, AY0031, site of Xiaomintun, China. Processed loess, levigated or decanted, with addition of lime (Stoltman *et al.* 2009). Photo: I. Druc.

90x

4.9 Ware made with clay and loess, resulting in a fine paste with only few fine-size inclusions. Mallard Bay #16-65, plain body shard, Cameron County, Southern Louisiana, USA. Cut ceramic fragment. Photo: I. Druc. The aligned texture and difference in paste color may come from the incomplete mixing of two raw materials.

4.6 to 4.9: Collections of the University of Wisconsin-Madison, Jim Stoltman. For production of ceramics with loess, see Stoltman *et al.* 2009, and Stoltman 2016.

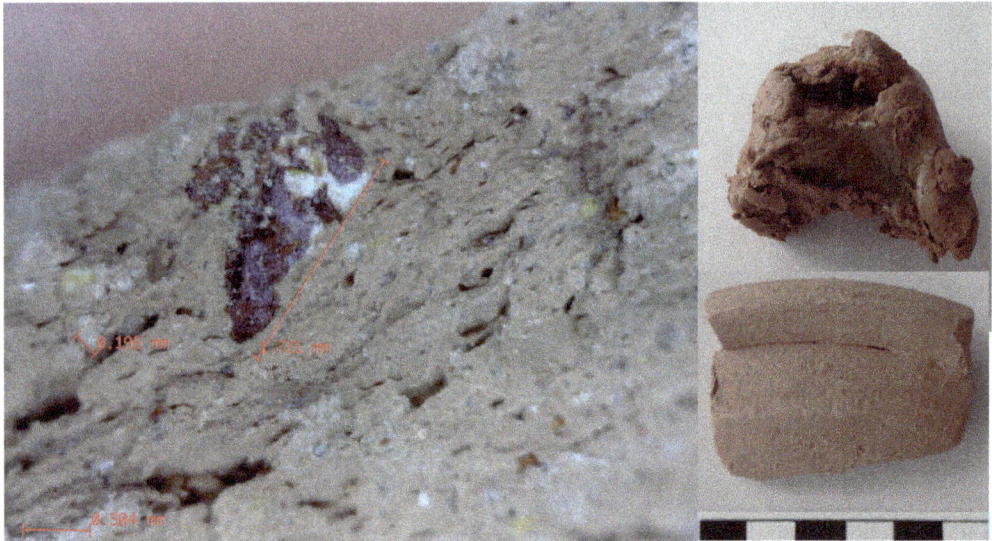

a) 95x (paste at left)

b) 150x

4.10 Untempered small stoneware cooking pot: (a) unfired paste, clay lump, and unfired ceramic fragment; (b) fired paste and fragment of fired pot at c. 1200 °C. Silty clay naturally rich in quartz, feldspar, and mica (and illite, kaolinite, chlorite and montmorillonite, Kirov and Truc 2012). Phu Lang production village, Vietnam. The clay is described as 'red'; note the white and red colors of the clay before homogenizing the clay ball. It fires red in an oxidizing atmosphere (exterior surface), but paste color varies from brick red to dark brown according to the type and amount of impurities in the clay and firing atmosphere.

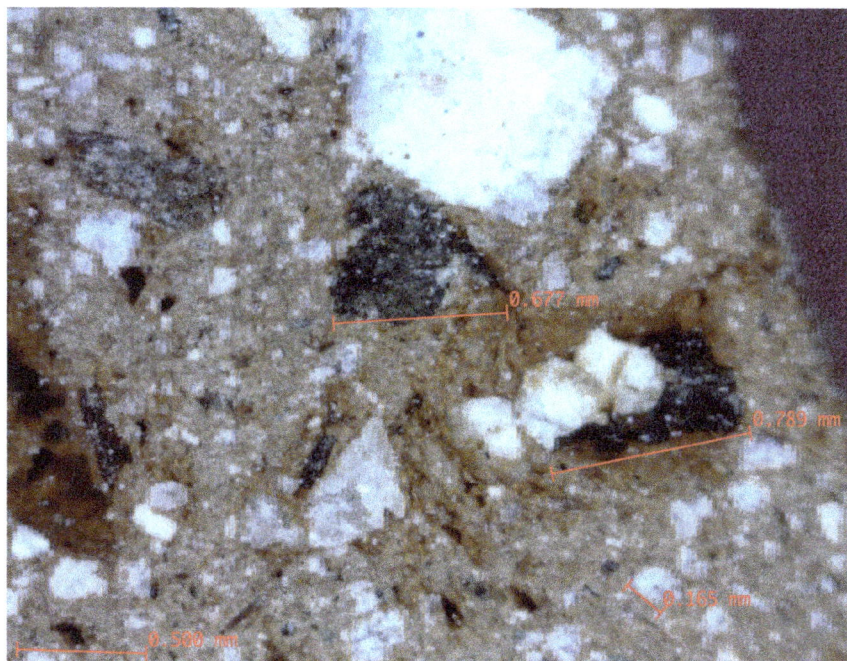

4.11 Untempered paste of a cooking pot, traditional production in Calpoc, Peru. 150x. The clay used is coarse, with abundant non-plastic inclusions of fine to coarse sand-size. We can distinguish many felsic minerals and igneous intrusive rock fragments.

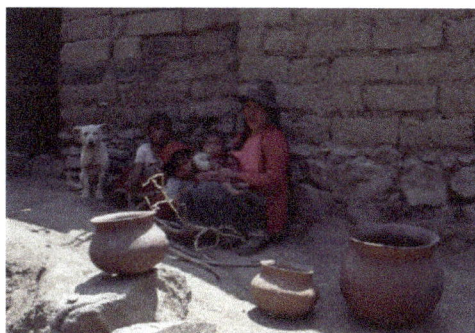

The angularity of the fragments indicates that the material probably came from an eluvial sediment (not transported far from its original deposit place). The place of production and resource area are located in the upper Casma Valley in the Black Cordillera, which is part of the Peruvian coastal batholith. Intrusive lithoclasts in the raw material reflect this situation.

In other cases, when the sediment or sand is transported by wind or water, the minerals suffer a series of physico-chemical processes, which impart certain sphericity, altered texture, dissolution and sorted grain-size. Robert Folk in his manual on sedimentary rocks (1965) details the characteristics of different sediments and their constituents. This information is very useful for provenance analysis of the raw materials used by the ancient potters.

4.2 Tempers

Sand temper

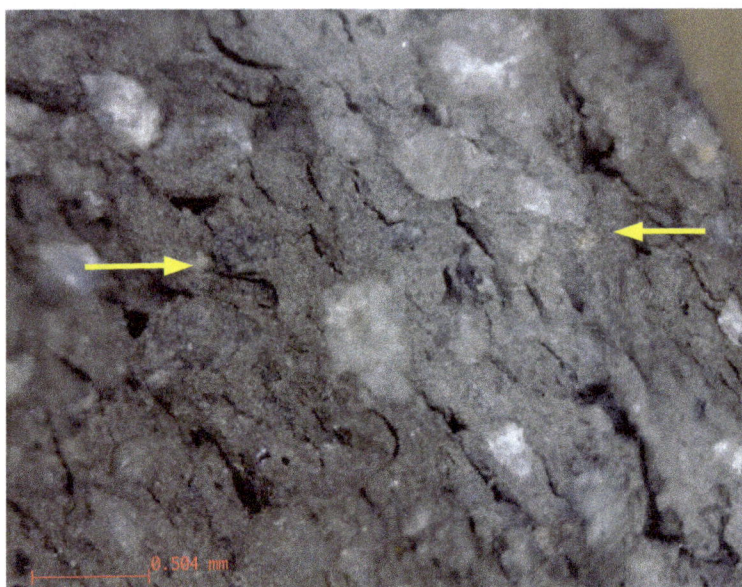

155x

4.12a Paste produced with clay and coastal sand as temper. The grains are subangular to subround, typical of a littoral environment. There are quartz and altered feldspar grains, volcanic clasts, mafic minerals, carbonates and fine recrystallized bioclasts or microorganisms. These can only be identified in petrographic thin sections, where the light-beige concretions (yellow arrows in the photographs 4.12a and b) can actually be identified. The black voids presenting half-moon shapes impart a particular texture to this paste. They may result from the dissolution or disappearance of organic material upon firing of the ware. Plate or disc PUCA50, Puémape, Peru.

PUCA50

0   5cm

4.12b 155x.

Grog temper

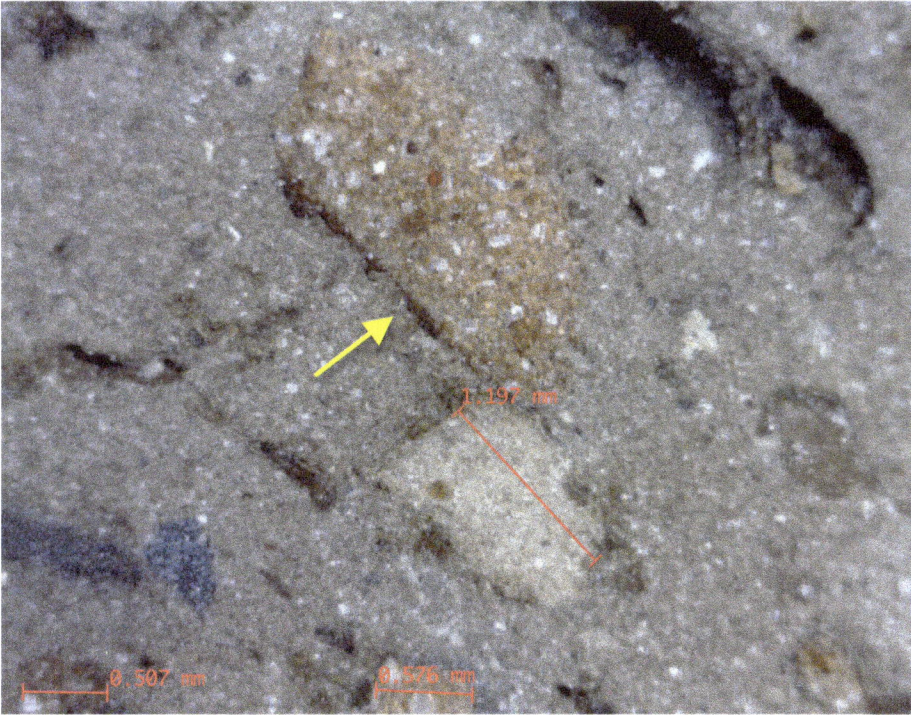

4.13 Grog-tempered ware, Mallard Bay #16-65, plain body ceramic fragment, Cameron County, Southern Louisiana, USA. Collection Jim Stoltman. Deparment of Anthropology, University of Wisconsin-Madison. Cut ceramic fragment. Photo I. Druc. 90x.

One side of the grog fragment above (light orange color) still displays remains of slip (yellow arrow). The angularity of this type of fragments in the paste and difference in composition and/or amount of small inclusions within the fragment as compared to the inclusions in the paste are characteristics to look for when trying to identify grog. The color and composition of grog fragments may vary from one fragment to the other, as fragments of different vases could have been crushed and incorporated into the paste. Another characteristic is the frequent paste retraction around a grog fragment (see Whitbread 1986 to differentiate between clay pellets and grog).

4.14 Grog-tempered ware, Jonathan Creek 15ML4C 90x (above) and 115x (below). Mississippian culture. Ceramic fragment from the collection William S. Webb, Museum of Anthropology, University of Kentucky, on loan to Sissel Schroeder, Department of Anthropology, University of Wisconsin-Madison. The vessel illustrated below belongs to the collections of the Department of Anthropology, University of Wisconsin-Madison. Photo I. Druc. See Schroeder 2009 for details on Jonathan Creek.

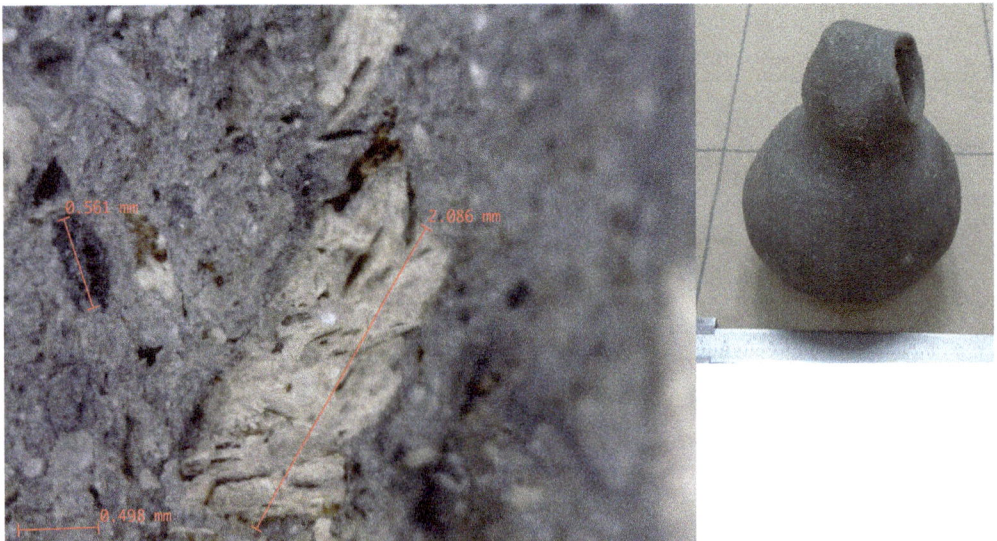

Shell temper

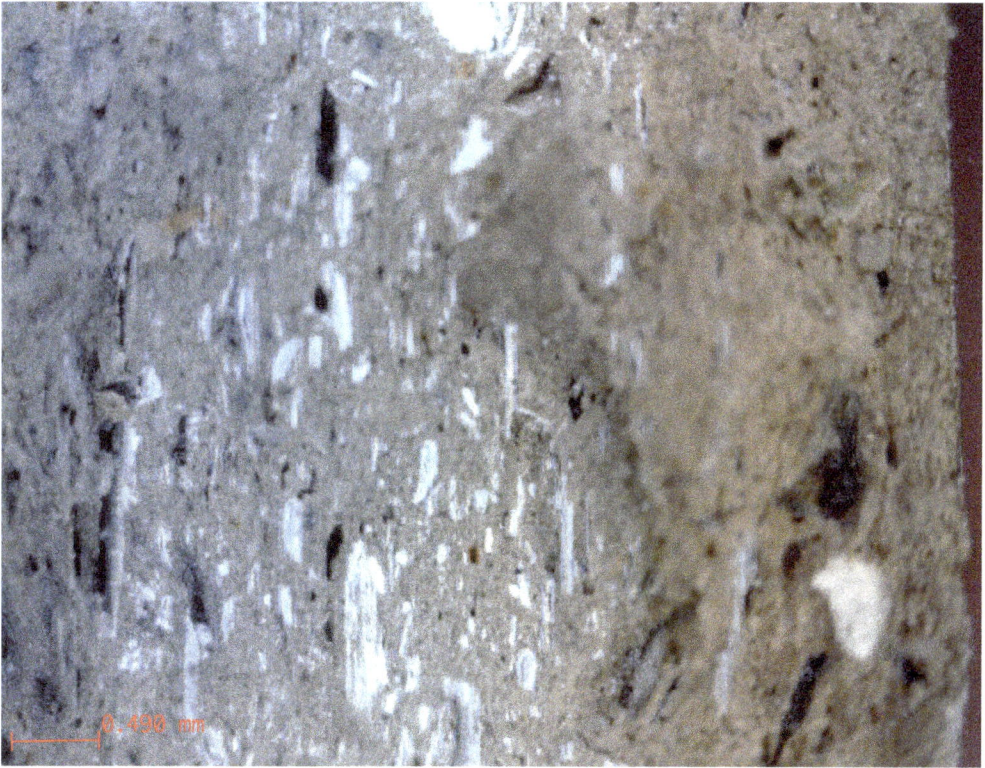

4.15 Shell-tempered globular base jar with short outward angled rim, Mississippian culture, Illinois. 90x. Ceramic fragment from the collection William S. Webb Museum of Anthropology, University of Kentucky, on loan to Sissel Schroeder, Department of Anthropology, University of Wisconsin-Madison. Note the alignment of the elongated shell fragments, their striation, and their white to grey color.

Crushed rock temper

Certain rock types are easier to grind than others, this is the case for low-grade metamorphic and sedimentary rocks like slate, shale, mudstones, or pelites (clay-rich clastic rocks). For other rocks, in particular those rich in silica, it is possible to expose them to the fire, which will render them more friable. The addition of crushed rock to a ceramic paste can be supected if a majority of the lithoclasts present are of the same rock type, rather angular and of similar granulometry.

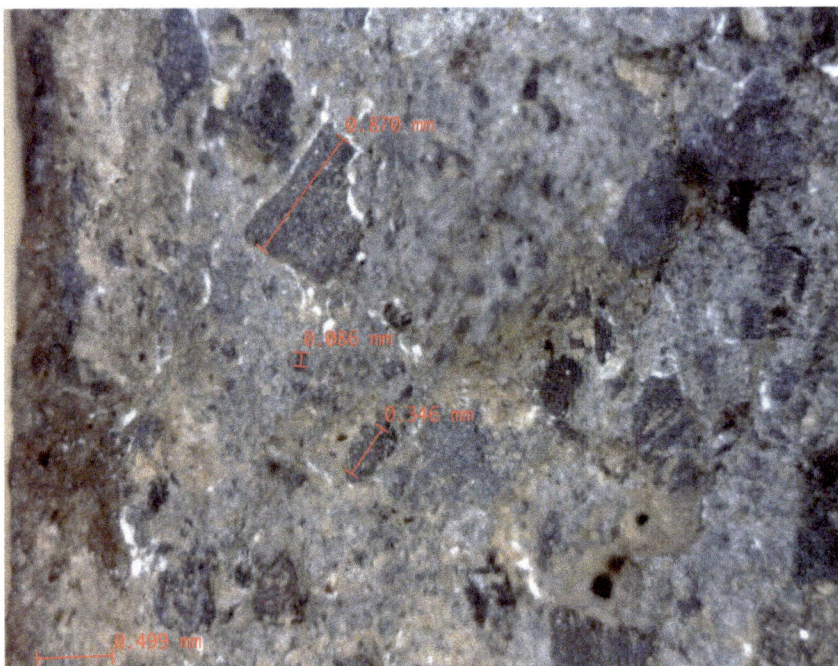

4.16 Archaeological ceramic tempered with fine to coarse sand-size crushed slate fragments (dark grey to black tabular metamorphic grains) with foliated structure. The paste was not well mixed and presents uneven grain distribution and orientation. MAR2, cooking pot, Marcajirca, Peru. 90x. This type of temper is still used in the local traditional production. Below is an example of a cooking pot from Mallas, Peru, produced with crushed slate temper and the profile of a jar neck fragment where the slate temper can be seen with the naked eye.

According to the region, shale is used as temper (a sedimentary rock that crumbles easily. Shale grades into slate.

Chaff temper

In the next two images, the vessels have been tempered with chaff. Upon firing, the organic material burned off, leaving only elongated voids (black). The wares are also rich in calcite, present both as primary (white) and secondary (grey) concretions. Secondary calcite fills out voids under certain conditions, espousing the shape of the original grain. The wares have been incompletely fired, and the core of the ceramic has not been oxidized, leaving a grey center.

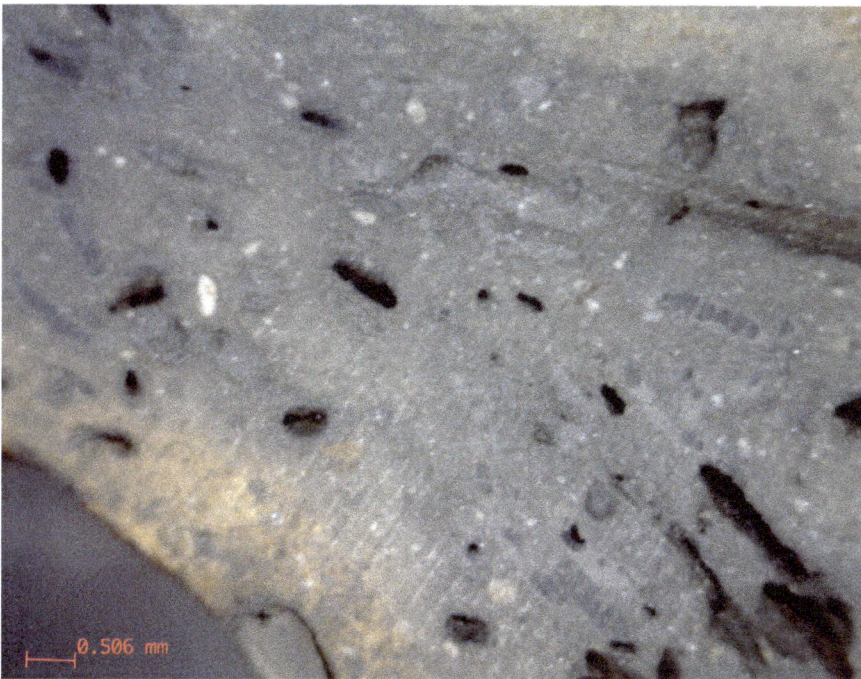

4.17 Imprints of chaff and calcite inclusions (primary and secondary). 50x. Bowl of the Dalma Period, Late Neolithic, Qaleh Paswah 5 Iran, Frank Hole Ceramic Collection, Yale University, USA. See the petrographic study by Yukiko Tonoike (2012, 2013) for analysis details. Photo: I. Druc 2013. Drawing: Y. Tonoike.

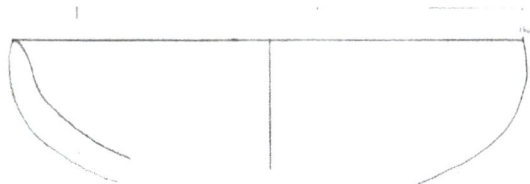

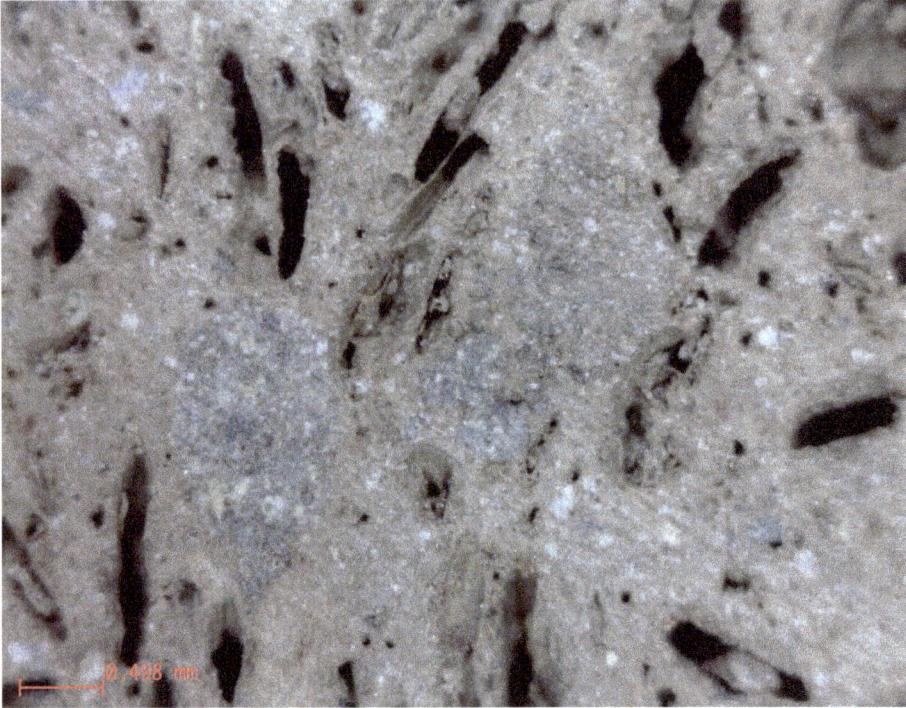

4.18 Ware tempered with chaff (seen as elongated voids after firing). 100x. Also visible in the photograph are subround silt pellets (grey rock fragments) with fine silt-size minerals. Dalma Period, Giyan, Iran. Frank Hole Ceramic Collection, Yale University, USA. See the petrographic study by Yukiko Tonoike (2012, 2013) for analysis details. Photo: I. Druc 2013. Drawing of illustrated ware: Y. Tonoike.

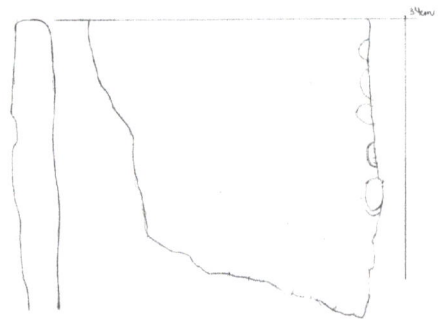

4.3 Manufacturing Techniques

The alignment of inclusions and pores in the paste points to the type of technique used to build a pot. As a potter may combine various building techniques for different parts of the ware, it is important to specify which part is being analyzed, unless we can assume the same technique was employed throughout the elaboration process. It is also possible that two techniques were used for elaborating the same part, like observed in Phu Lang, Vietnam, where the walls of a jar are built piling coils that are shaped with the wheel. Below are illustrations of manufacture with the coiling technique (figure 4.19) and the paddle-and-anvil technique (figure 4.20). See Rice (1987) for a summary of manufacturing techniques to help with data interpretation.

4.19 Anaseta Ocaña Janampa building a large jar with the coiling technique. Yacya, Ancash, Peru. 1997.

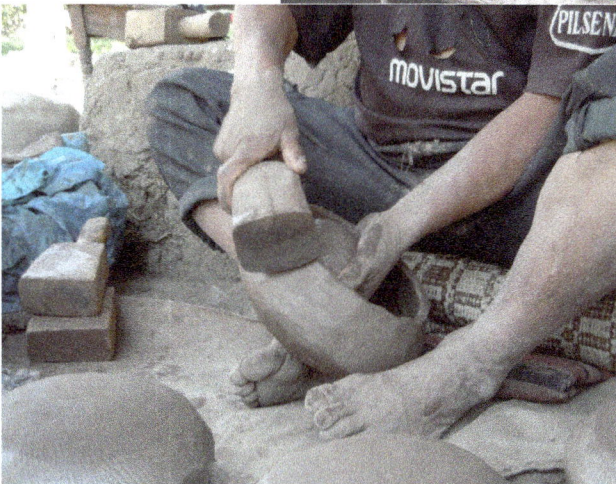

4.20 Paddle-and-anvil technique, Miguel Tanta Aguilar, Mangallpa, Cajamarca, Peru. 2010.

4.21 The curvilinear alignment of the grains and fissures in the paste indicate the presence of a coil, attesting that part ot the pot was built with the coiling technique. Bowl, upper part, volcanic paste, ID12, Kuntur Wasi, Peru. 100x.

4.22 The alignment of the finer crystals and pores in a circular fashion suggests the use of the coiling technique. This is seen in the upper part of the vessel and the coil(s) would have been 1.6 to 2 cm thick. However, the elaboration process might have involved another technique or the compression of the walls in the lower part of the vessel. There, the wall is thinner and the pores and inclusions are parallel to the surface of the ware (photograph below). A paddle could have been used to compact the walls, and stretch or reinforce them. Bottle body KW3, Kuntur Wasi, Peru. 150x/90x.

4.23 The presence of a coil is only noted by the bulge and curve in the paste toward the exterior part of the vessel, while the internal face has been flattened. Apparently the interior and exterior surfaces did not receive the same amount of pressure when strengthening the walls. Bowl CP35, Kuntur Wasi, Peru. 90x.

4.24   Parallel alignment of pores and inclusions suggests pressure on the sides to form the vase. (e.g. with a paddle, or pressuring into a mold). Bowl KW64p, Kuntur Wasi, Peru. 85x.

4.4 Paste Textures

The term texture in ceramic studies describes the general aspect of a paste consisting of the distribution, size, percentage of inclusions and pores or voids, proportion of clay matrix to non-plastic grains, and appearance of the clay matrix. The later is best seen in petrography.

Ceramic pastes are often categorized as fine, medium, or coarse, and technological analysis may look for correlations between the granulometry of a ware and the form or style produced. Here are a few examples of fine to coarse wares related to earthenware productions. More discussion about paste granulometry, analysis and classification is given in Chapter 5. Interpretation of the granulometric data to evaluate the level of technology, knowledge or expertise of ancient potters should be made with caution. Different types of vessels, ware functions and raw materials used, or even tradition, may require working with more or less refined materials. Stoneware and porcelain productions present differences in the type of clay used, which must withstand high firing temperatures (between 1100 and 1600 °C according to the type of kiln and placement of the product in the kiln). High firing and silica-rich compositions contribute to the strength and impermeability of stoneware and porcelain products, even if the granulometry of the inclusions is fine (see figure 4.28). For earthenware products, temperatures are lower (usually between 700 and 900 °C) and granulometry can vary more than for stoneware products. The use of coarse material in earthenware production yields strength to a ware, and is needed for the construction of large jars or cooking pots.

The fine or coarse character of a paste and its texture depend upon the vase to be produced, the size and function of the piece, the components used and the technological tradition to which the potter belongs. Coarse pastes do not necessarily imply bad or inferior technique, or haste in execution. On the contrary, it can point to the experience of the potter and his or her knowledge of the prerequisites necessary to produce a pot that will withstand wear, thermal shock or transport. Although a very fine paste is often characteristic of fine vessels, bottles and wheel-thrown pots, this is not always the case. Note that all the vessels illustrated in this manual are hand-made, except the stoneware pots from Phu Lang which are wheel-made (figs 4.10b, 4.28, 4.33).

Very fine and compact paste

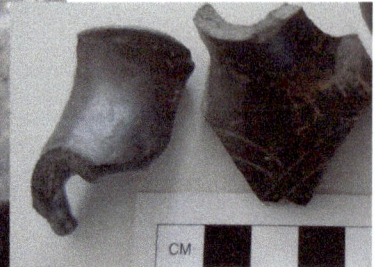

80x

ID42

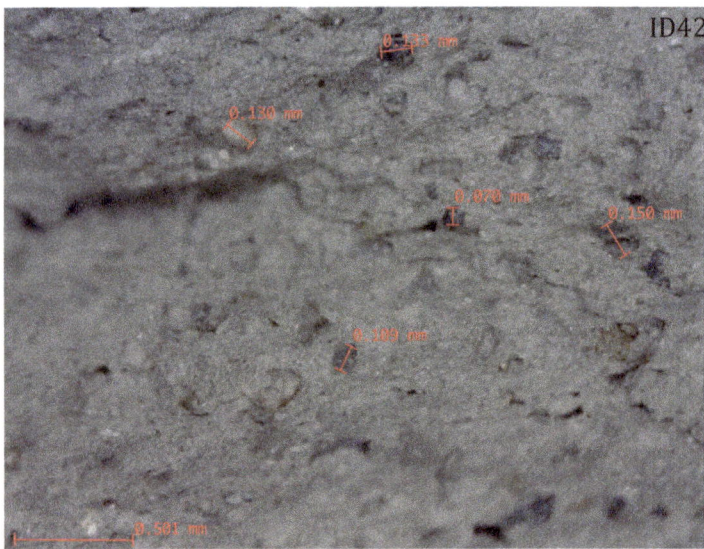

165x

4.25 The unimodal (one size) granulometry and similar composition of the inclusions suggest that the potter used one material and probably refined it. Refining could have been done by sieving, decantation or levigation to eliminate the medium and coarse fractions present in the original raw material. Homogeneous distribution of the inclusions, lack (or small amount) of pores, and compaction of the paste indicate a good level of working (kneading) the paste before use. The clay matrix presents a microgranular texture. Bottle ID42. Kuntur Wasi, Peru.

Medium paste and controlled granulometry

4.26 Elimination of the coarse sand fraction by sieving and/or decantation of the raw material(s). Good kneading is attested by the homogeneous distribution of the grains. The lack of pores may be related to kneading, firing and the type of material used. Paste compaction may relate to a molding technique in the

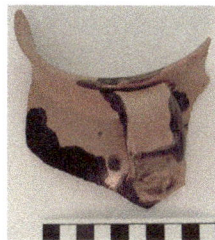

elaboration of the wares. Mineral composition consists of quartz and plagioclase crystals (white subangular to subround grains), iron oxides (reddish brown) and mafic minerals (fine brown to dark prismatic and elongated grains). Cooking pot, Toluca, Mexico. 85x.

Coarse paste, mixed composition and granulometry

neware body

4.27 Coarse paste with moderately sorted materials, no grain orientation, and inhomogeneous distribution of the inclusions (either close to the surface or at the core). A few large elongated voids are also noticeable. Traditional cooking pot, Marcará, Peru. 90x.

Sto

4.28 Stoneware jar fragment, untempered silty clay. 165x. High firing temperature (c. 1200 °C) and elevated contents of quartz, feldspar, illite and kaolinite produced a dense and nonporous material, characteristic of stoneware bodies. Smaller pore sizes reduce porosity and increase impermeability. The range of colors in the glaze is due to impurities and thickness of the glaze. Here the interior surface is glazed for impermeability, as these jars are used for storing fish sauce; the exterior surface is left bare. Note the typical 'hard' surface aspect of stoneware products. Jar elaborated from thick coils shaped with the wheel. Traditional production, Phu Lang, Vietnam.

4.5 Slips and Glazes

A slip is a mixture of clay and water applied onto the surface of a ware. The slip may present the same mineral composition as the clay matrix if produced with the same clay, but the inclusions, if any, are usually very fine. It can be burnished and this action will produce an alignment of the clay particles parallel to the surface giving it a shine. The presence of a slip or paint should show as a distinct layer with a clear limit with the paste underneath, with no diffusion of color or material towards the center of the ceramic. Note that a surface can be burnished and produce a shine without a slip and that the two surface treatments can be difficult to differentiate in some instances. However, the burnishing, although it brings up and aligns the clay particles, will not present a layer as does a slip. For more information on slips and burnishing from a potter's perspective see Von Dassow (2009).

A glaze is made of silica-rich materials such as found in quartz and feldspar (for the glassy look) and alumina (in feldspars) to harden the glaze. The silica-rich components are mixed to a substrate, and the mixture can be applied dry or wet. The impurities or oxides added or naturally present in the raw materials color the glaze. The red or black glazes of Roman sigillata ware, for example, are produced with much the same clays as in figure 4.33 for the green glaze but with different organics. They produce a black glass when the iron is reduced to a magnetite state, or without organics, an oxidized red glass surface. High temperatures (usually above 1100 °C) are required to melt the glaze over the ceramic body, unless fluxes are used to lower the melting point. This produces an impermeable, semi-vitreous or vitreous surface, that can be opaque or not. It is the potassium in the illite in the clay (or in the added leaf ash as in the example in figure 4.33) that helps fuse the clays. The glass state resists changes in furnace oxidation or reduction after it is formed, such as in Greek black on red ceramics (see Velde and Druc 1999). Glazes are sometimes called enamels, however true enamels should have powdered glass mixed to the coating substrate.

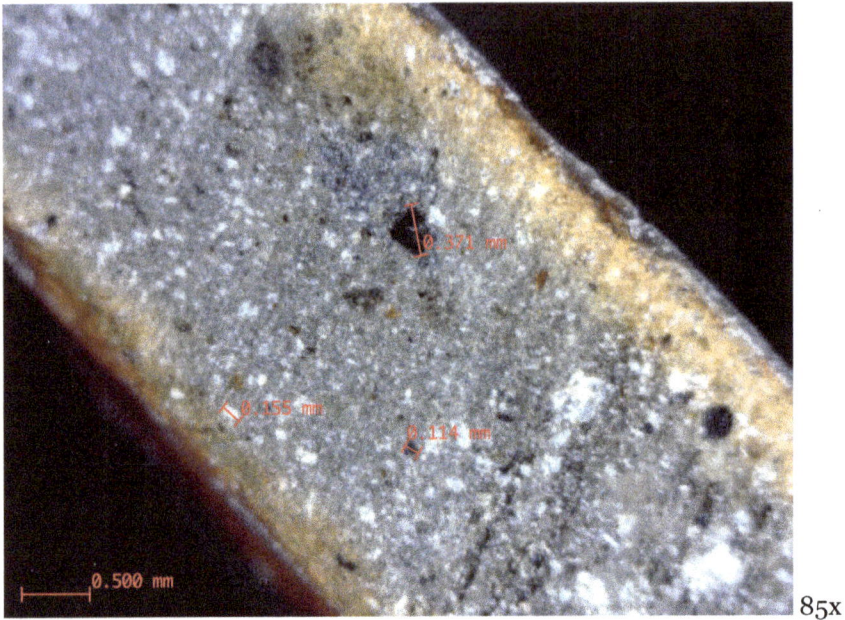

85x

4.29 A fine reddish slip, possibly rich in hematite, is readily seen on both sides of the cross-section of this ceramic. The white film at the exterior surface of the ware (seen as a white line in cross-section, to the right of the photograph) is a post-use deposit. The ware was fired in a reducing atmosphere with oxidation only happening at the end of the firing (light brown paste color with diffuse limit with the more reduced center), probably during the cooling period. The paste composition mainly consists of fine-size quartz and feldspar crystals, and lower amount of felsic (silica-rich) rock fragments of unidentifiable composition, as well as a few iron-rich grains (orange) and mafic minerals (dark inclusions). Petrographic analysis revealed that the rock fragments were of acid to intermediate intrusive composition (granite to granodiorite).

Open bowl CP65, Kuntur Wasi, Peru. Saw cut ceramic fragment.

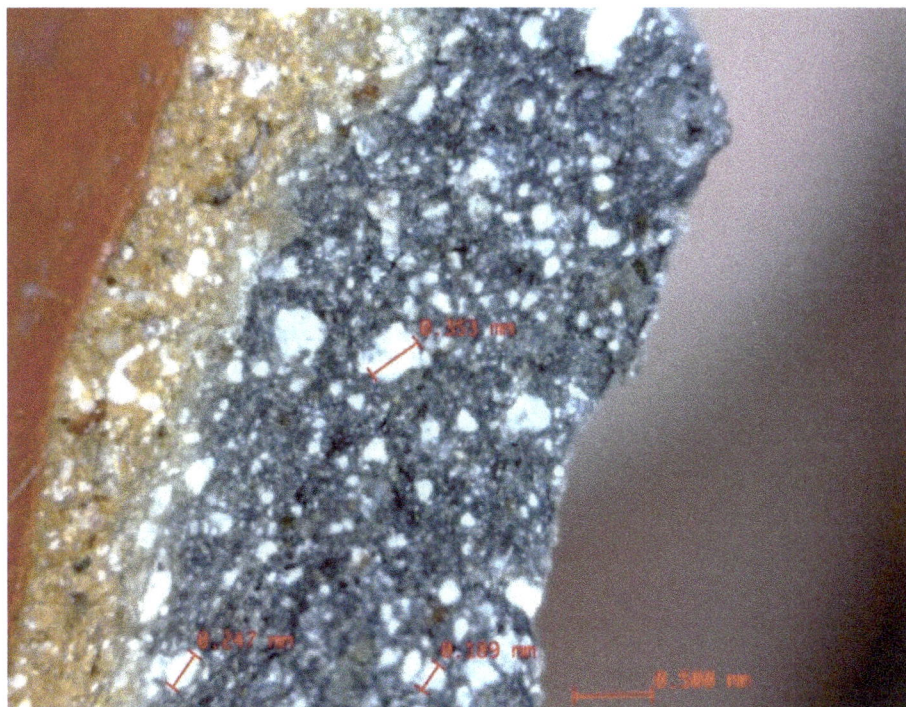

4.30 Fine red slip coating the exterior side of the ceramic. Note that the oxidized, underslip fringe (light brown) presents the same composition as the interior of the ceramic. The difference of color is only due to firing and differential oxygen access between exterior and interior sides. On the contrary, the thin slip at the surface of the ware has a very fine granulometry and is very dense. The limit between the slip and the paste body is precise. Bowl KW66bp, Kuntur Wasi, Peru 85x.

4.31 Burnished bottle spout without slip. Reduced clay body with superficial oxidation on the exterior surface of the bottle. Reduction is due to restricted oxygen access to the interior. Bottle CP84p, Kuntur Wasi, Peru. 85x.

4.32 The very fine black exterior layer may result from intentional carbon smudging of the surface at the last stage of firing, or from the addition of graphite or a manganese-rich slip. Surface analysis with Raman spectroscopy or scanning electron microscopy can confirm the nature of the slip. Bottle spout KW94p, Kuntur Wasi, Peru. 65x.

4.33 Quartz and feldspar-rich glaze (green) on the outside surface (right) of a large stoneware jar from Phu Lang, Vietnam. 80x. The high temperature reached during firing (circa 1200 °C) hardens the body to a stoneware quality, and melts and vitrifies the glaze. Here, a mix of river mud and bamboo leaf ash was used.

The jar on the right is glazed on the interior for impermeability and exterior with a different plant ash to obtain a chocolate brown glaze color for decoration. Below: unfired glazed jars. Phu Lang, Vietnam.

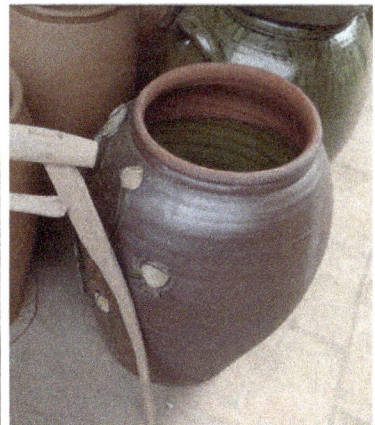

4.6 Firing

Many variables affect the end-result of a firing and final color of a clay body and surface of a ware. An oxidizing atmosphere usually produces a ceramic with reddish or light brown tones according to the amount of iron oxides in the clay. However each clay fires to different tones and responds to the firing atmosphere differently. Furthermore, color differences are often witnessed in cross-section between the center and the sides of the ceramic. This can be due to an incomplete oxidation or incomplete burning out of the organic material, which is often present naturally in many clays. This depends upon the firing temperature and atmosphere, quantity of oxygen available, porosity and composition of the paste, granulometry, length of heating cycles, etc. There are clays that lack organic material but are already oxidized. In this case, a firing in an oxidizing atmosphere will not affect the color of the paste, but it will change if fired in a reducing atmosphere to yield a light to reddish center and a dark surface. To obtain true reducing conditions, oxygen access must be totally blocked to allow for the chemical reactions to take place, transforming the organic material and iron present in the paste. See von Dassow (2009) for firing conditions as well as Rye (1981) for the interpretation of the paste color in relation to firing or Velde and Druc (1999: 122-128) for an explanation of the effects of oxido-reduction and resulting chemical changes.

4.34   Incomplete oxidation of the organic material, which left a black core. Cooking pot KW18, Kuntur Wasi, Peru. 80x. Firing in an oxidizing atmosphere.

4.35 Paste colors result here from a restricted oxygen access to the interior of the bottle during firing. The body of the ceramic is half oxidized (on the exterior side) and the surface is dark brown and very well polished. The color difference seen in the upper left part of the cross-section is due to surface deposit occulting the paste. The fresh break did not include that part. Bottle body with incised decoration, Pallka, Peru. 100x.

4.36 Firing in an oxidizing atmosphere. The thick black layer on the surface (red arrow, left) is not due to the firing of the pot but to carbon deposit upon recurrent use for cooking. Traditional cooking pot, Musho, Peru. 90x.

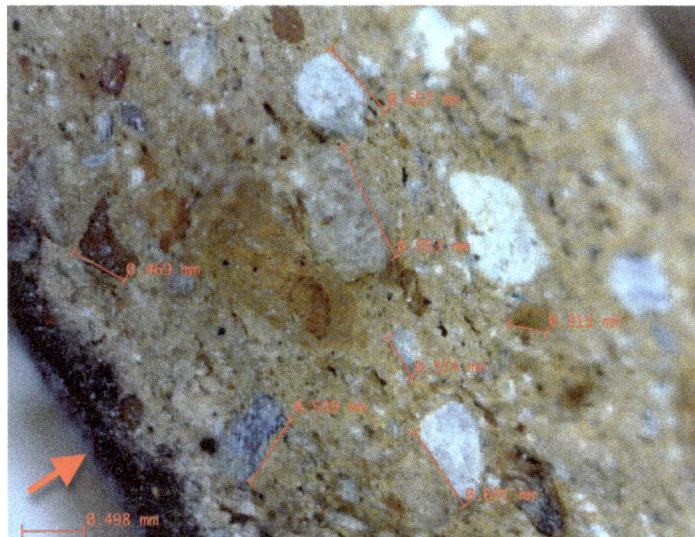

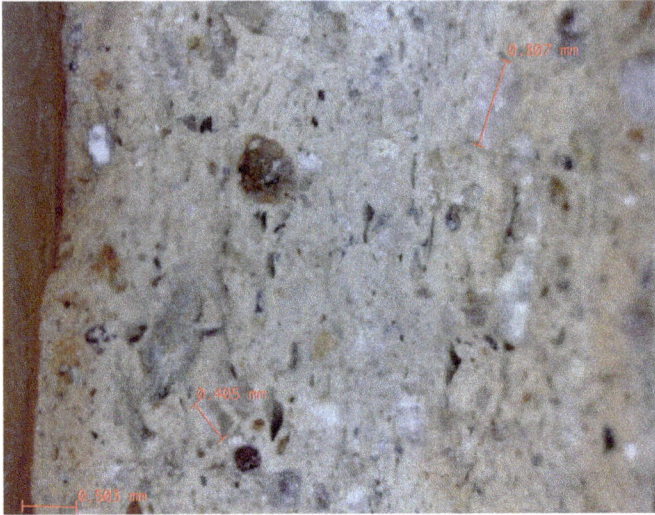

4.37 Firing in a wood kiln, traditional production, sample CA7, Taller Manya, Cajamarca, Peru. 90x. Even if the ceramic has been fired in a closed kiln, the paste is still partly oxidized. This is due to the high temperature reached, the type of clay used and oxygen available in the firing chamber. The light-colored center of this paste indicates that the raw materials used were poor in iron or organic material. The paste is rich in quartz and plagioclase crystals, with minor presence of biotite, hornblende, and quartzite fragments. A slip is visible on the exterior surface to the left.

4.38 Neutral firing, without excess or lack of oxygen. This produces a uniform color of the paste. Bottle body KW92p, Kuntur Wasi, Peru. 85x. Below: photograph of the interior of the bottle.

5. IMAGE ANALYSIS

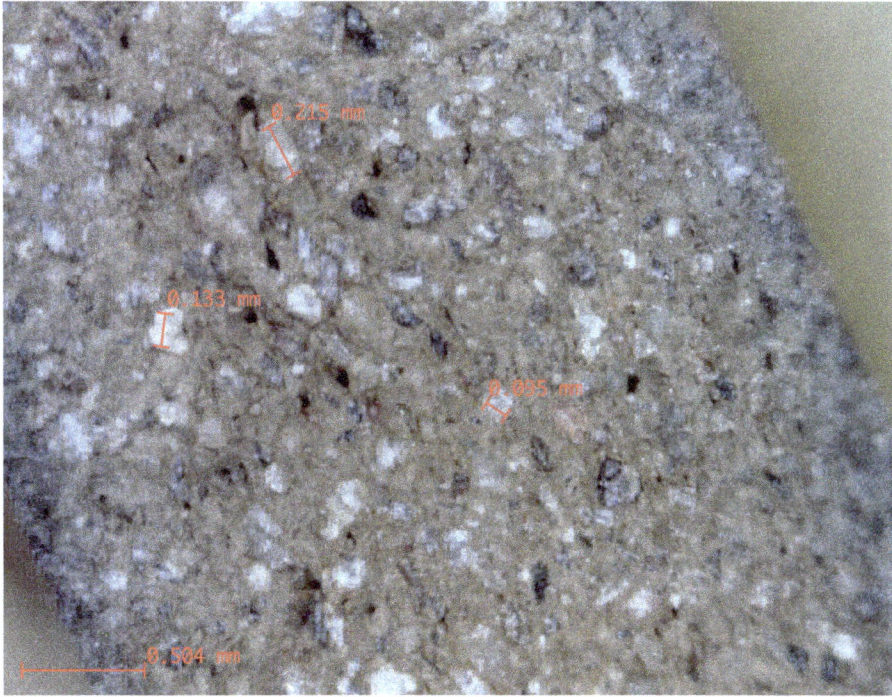

5.1 Group 1: Paste rich in subround crystals of quartz, feldspar (a, K), plagioclase and mafic minerals, with fine homogeneous granulometry, and nearly no lithoclasts. Bowl PU120, Puémape, Peru. 135x.

5.1 Image Analysis Protocol

One of the objectives of macroscopic paste analysis of a ceramic corpus is to discover recurrent mineral and textural traits that allow us to group ceramics with similar characteristics and identify those with atypical ones. One can later select ceramic fragments that would best represent a paste group to conduct a more detailed analysis, with petrography, scanning electron microscopy, neutron activation or other methods. Detailed mineral and/or chemical analysis helps confirm the preliminary grouping and describe paste composition. As stated at the beginning of this manual, several details cannot be observed with a binocular or digital microscope, nor can petrography or chemical analysis answer all our questions alone. It is best to use a combination of approaches. Nevertheless, the groups formed based on macroscopic analysis can give valuable information, providing there is some system and rigor in their analysis. They can be compared and results can be triangulated to stylistic, formal, stratigraphic, or contextual data for example, in order to initiate a reflection on technology, production, provenance, and distribution. Distinction of local from non local provenance, however, requires comparative and geological data, and other information considered necessary for the time period and region of study. It is useful to remember that ceramics are best compared among themselves or test tiles, rather than with raw clays and temper materials, and thus the addition of ceramics from other contemporaneous sites is very informative. The latter is particularly important because it gives an idea of the technological tradition and raw materials used in a region, providing that the comparative ceramics used are representative of the local tradition at that site.

The analysis protocol proposed here refers only to the preliminary steps of ceramic analysis, mainly the acquisition of macroscopic data of ceramics in hand specimens observed with a binocular or digital microscope. Such a protocol must be adapted to each archaeological problem and project.

As a first step, it is useful to quickly look with the microscope at a great quantity of ceramics to have a first idea of the paste variability in the ceramic corpus. Then classification criteria that will define the different paste groups can be proposed, such as granulometry, basic mineral characteristics (relative presence of felsic or mafic minerals, or rock fragments), grain angularity, etc. Ceramics are then grouped according to these defining criteria and visual characteristics. The groups are illustrated by images of representative pastes of each group, taken with the microscope to constitute a reference data bank. A description of the group is also given (see for example figure 5.1 and accompanying group description). This classification is refined as the analysis goes on, group homogeneity and consistency are re-evaluated, groups can be subdivided or new groups created when necessary. It is an iterative process that evolves with a deeper knowledge of the analysis corpus.

Decades ago, Matson (1970: 595) and Rye (1981: 50) proposed the creation of reference sets, with fragments of typical pastes (or temper types) glued onto a card to illustrate each group. The images taken with the digital microscope fulfill the same purpose and allow for archiving an important amount of data. Each group is illustrated by paste images, as many as necessary to show the range of internal variability. Group description should be clear so as to allow other analysts to work or continue the classification.

When the groups have been defined, group attribution of new fragments is rapid. Analysis can then proceed. One notes the type, style, form and any other contextual details deemed useful, the number of fragments per group, etc. Such a process is illustrated in detail in Marsh and Druc (2015).

In summary:

1. For each stratigraphic level, unit or context, look at the ceramic pastes with the digital microscope and regroup them according to their mineral and textural similarities. Break off with pliers a small fragment of the ceramic to obtain a fresh cut, that way surface deposits will not hinder identification. The digital image of the paste will be displayed on the computer screen via a USB connection.

2. Write down in an Excel table or other, the number of fragments per paste group, and fragments per style and/or form per paste group.

3. Take pictures with the microscope of the paste of one or more ceramic fragments in each paste group and, with a camera, of the fragment(s) studied (front and back). If there is compositional or textural variability within a group, document this by taking photographs illustrating the variants or exceptions, while specifying in your notes that these are variants of the paste group(s). Save the images in a separate file, in jpeg or tiff, with a scale at the bottom (e.g. 0.500 mm). Do not forget to note, when labeling the image for example, at which magnification the photograph has been taken.

4. With the image analysis program provided by the digital microscope measure a few grains of different dimensions (fine, medium, coarse, very coarse). This allows for a later, rapid estimate of the percent of grains in the paste per granulometric class. When drawing a line to measure a grain, try not to cover any diagnostic feature that would allow mineral identification if this is done on the image you are saving for your archives. One can trace a line at the side of the grain (see figure 5.2).

5. Note any impression, comment, similarities, and differences seen within and between groups. Your mind works better than a computer but your memory does not. This information can be useful for group description and intergroup comparison.

5.2 Example of image annotation. Measurements are taken along the grain or across, according to work modality. Cooking pot fragment, Sorkun, Turkey. 150x.

6. Select one or more fragments per compositional group, per ware forms and ceramic styles for thin section and petrographic analysis (or other). This allows seeing the differences in paste composition in relation to ware function or ceramic style. Also select any fragment with atypical or uncommon characteristics.

7. Draw a line on the ceramic fragment where you want the technician to cut for the thin section, avoiding diagnostic features. A profile cut (cross-section) is useful to get the sides to observe manufacture differences.

8. Separate the ceramic samples for detailed analysis (petrographic or other) in a plastic bag with complete ID, specifying bag number and other information useful to put the fragments back in their bag once the analysis is completed. The samples should be photographed (paste and surface). Leave a note in the bag from which the ceramic fragment was taken, writing down the identification number, form, style, date, and name of person who did the selection.

For petrographic thin sections, the ceramic fragment must be big enough (minimum 2 cm x 2 cm, or 3 cm x 1.5 cm). If possible it is best to choose a side for the cut devoid of diagnostic features. Certain tools and expertise are needed to make the cuts and prepare the thin sections, which includes epoxy impregnation and thinning the sample to reach the required 30 microns thickness. Although kits are now available to do the job in the field (Goren 2014), this part of the work is best left to a specialist or trained person. Similarly, petrographic analysis requires a particular training and the use of a polarizing microscope (where the light comes from beneath the sample), with high magnification and a rotating stage. These microscopes are great, but big, heavy, expensive, and not easy to transport to the field. Alternatively, you can buy a small backlight polarizer stage that can be coupled to the portable digital microscope, which should have the polarization feature if you want to conduct basic petrographic analysis in the field (see figure 1 at the beginning of the Atlas). This allows one to perform decent mineral identification while in the field, providing you have thin sections. However, the stage does not rotate -at least not like on a petrographic microscope-, which hinders some observations or measurements of the extinction angles, and the magnification of the portable microscope is often too low for a detailed analysis. No doubt improvements in portable devices will change fieldwork possibilities, but archaeological budgets still need to match the technological advances.

A basic image analysis program usually comes with any digital microscope, which allows you to put a scale, measure, comment, label, or draw on the image. Photographs are automatically saved in a folder linked to the digital microscope program, but it is preferable to save them independently in a separate folder. Thus, if the link between the photograph with its identification and your notes is lost, you can easily find the data. A tiff format offers better resolution than a jpeg one, but it 'weights' more (e.g. 35 MB in tiff format vs. 3 MB in jpeg). PNG is a good format too.

For publication, you need an image with at least 300 dpi (dot per inch). If the default setting of your digital microscope program is lower, you will have to adjust the resolution 'manually'. The image size in the program Dino-Lite, for example, is 1200 x 960 pixels, but resolution is at 72 dpi. To change this, you can open the image in another program (Preview, Photoshop, Gimp, etc), and augment the dpi, while reducing the size of the image.

5.2 Quantitative Image Analysis

Digital microscopes now come with basic image analysis programs, with measuring and annotation tools. Usually, they do not offer quantitative image analysis but many programs of that nature are available through the Internet. Programs for geology, and sedimentology in particular, are the best adapted to the problematic of ceramic paste analysis. For example, JMicrovision is a free program created by Nicolas Roduit (www.jmicrovision.com). As required by professional and academic ethics you have to cite the source and author of the program you will use. These programs allow counting grains per size and mineral type, measuring angularity or the total area of grains in the clay matrix. These are semi automatic processes, as the analyst intervenes in all decision making steps, be it for identifying which grains need to be counted and how, or the categories to measure. Sometimes the analyst will have to identify by hand the grains to take into consideration or even swiftly draw a line across the grain to measure. The data is then automatically tabulated in an Excel file or other.

The reason for the lack of total automation lies in the limited color contrast between certain grains and the clay matrix of a ceramic paste. Felsic (white) minerals usually stand in sharp contrast against the tan, brown or dark clay matrix, and they can be counted by a program based on color contrast between the object to measure and the background. Mafic (brown, dark) minerals, on the contrary, present colors similar or with low contrast against the clay matrix, unless the color of the fired body is clear. The most salient grains will be counted, the others not, leading to a biased count. It is often easier and faster to measure manually 200-300 grains. One just draws a line across the grain and the program measures its dimension and registers the data in a table.

Quantitative analysis allows estimation of the percentage of certain grain types and performing objective comparisons between ceramics or paste groups. Many articles present such a methodology applied to ceramic analysis in thin section or to images taken from saw-cut ceramic profiles (e.g. Livingood and Cordell 2009, 2016; Middleton *et al.* 1991; Reedy 2006; Velde and Druc 1999; Whitbread 1991). Another article of reference is by Middleton, Freestone and Leese (1985) who discuss the effects of sampling upon grain-size distribution and present an image analysis program helpful for textural analysis.

5.3 Case Study

The example below briefly illustrates the analysis that can be done with paste images taken with a digital microscope. The analysis and histograms were performed with the program JMicrovision (Roduit 2002-2008 www.jmicrovision.com). The measures of the minerals and lithoclasts are given in millimeters (X axis). The Y axis gives the number of same-size grains.

The two histograms below (figures 5.3a and 5.4) show the granulometric range seen in two ceramics from the site of Kuntur Wasi in Peru, based in the measure of some 200 grains within the area photographed. The program allows moving along the image by set intervals, automatically or manually, following an axis, randomly or within a given area, and measuring the grain at each stop point. One can ask the program to perform a modal analysis, specifying what to do and entering the variables to count or measure. Grains can be counted or measured according to composition (quartz, plagioclases, micas, lithics, etc.) and granulometric sizes. Modal analysis takes time and is justified in petrography where it is easier to identify mineral compositions. It is however possible to do this with hand specimens, but on a more restricted scale using broader categories (e.g. felsic, mafic, lithics, fine, medium, coarse – see example in figure 5.3b).

The paste photographs next to the histograms were taken at 160x and the annotations (measures and scale) were done with the Dino-Lite program. These measures served to calibrate the measuring tool of the granulometric analysis program. The paste image was imported from the Dino folder where it was stored to the JMicrovision program (Roduit 2002-2008) to perform the analysis. Note that a fine, compositionally homogeneous paste requires fewer measurements than a coarser and more heterogeneous paste to obtain a representative vision of paste granulometry.

5.3a Histogram ID10, bowl, Kuntur Wasi, Peru. Area analyzed 17.45 mm². Medium paste. Analysis performed with the program JMicrovision (Roduit 2002-2008).

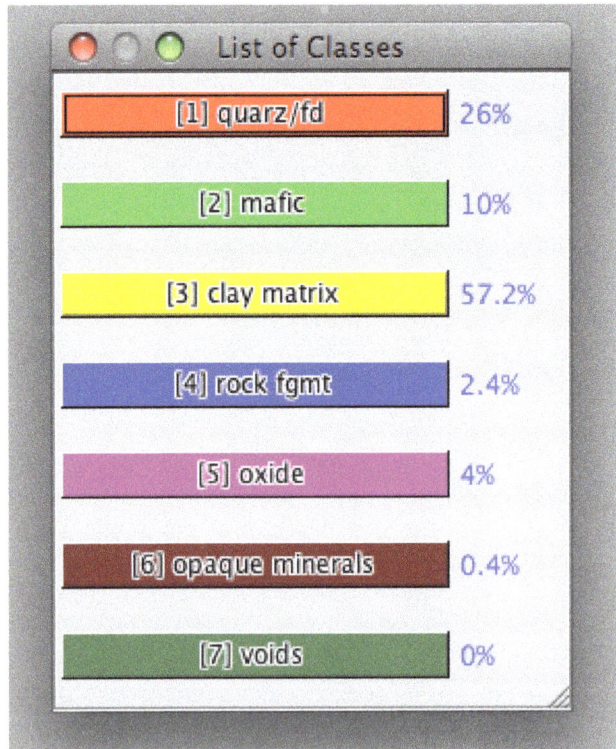

5.3b Result of the modal analysis with 250 grains counted (point counting, JMicrovision program, Roduit 2002-2008).

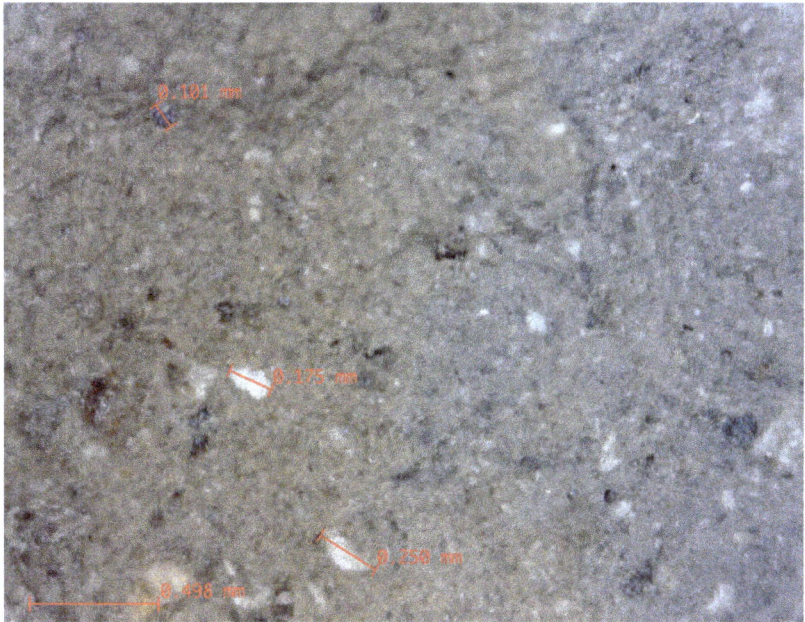

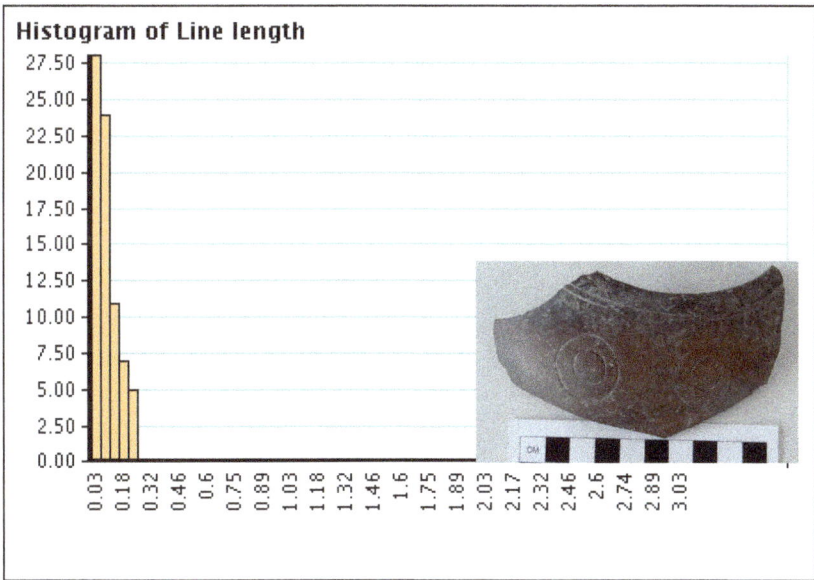

5.4 Histogram CP4, bowl, Kuntur Wasi, Peru. Area analyzed 20.06 mm². Fine paste. Analysis performed with the program JMicrovision (Roduit 2002-2008).

APPENDIX
Archaeological sites and places of ceramic production mentioned in the text, and relevant bibliography

Acopalca/Mallas/Yacya

Three neighboring highland villages of the Conchucos region, Ancash Department, Peru, where ceramic production used to be practiced on a large scale. Now only a few families are still engaged in the trade. The women are the potters while the men are in charge of acquiring the material and a male specialist fires the wares. The same clay and slate (temper) resources are often mined by potters of the different villages. The coiling technique is used. Druc 1996, 2005.

Ancón

Archaeological complex on the central coast of Peru. The site was identified as an important necropolis and a fishing village, with occupation between 1800 and 300 B.C. Druc *et al.* 2001; Rosas La Noire 2007; Willey and Corbett 1954.

Cajamarca (city)

Capital of the eponym department, the town of Cajamarca is located in the northern Peruvian Andes at 2700 meters above sea level. There are two main ceramic production areas within the city, mostly small family-operated workshops: the Mollepampa neighborhood where utilitarian ceramics are produced with bivalve molds, and the area of Cruz Blanca to the north of the city with artisans specializing in the production of decorated tourist ceramics and some utilitarian wares. Production in Mollepampa used to be done with the paddle-and-anvil technique until the 1980s. Druc 2011.

Calpoc/Cunca

Two hamlets in the Sechin and Casma Valleys on the central coast of Peru, where a few potters still produce on an occasional basis. Cunca is in the lower Sechin Valley. Production is done with clay from a nearby communal water canal and river sand is used as temper. Calpoc is in the upper Casma Valley. The only potter producing there uses coarse untempered clay to produce utilitarian ware. Both regions are in the Ancash Department. Druc 1996.

Cancharumi, see *Marcará*

Giyan

Iranian site with remains dating from the Late Neolithic until the first Babylonian dynasties (middle of the 5th until the 1st millennium B.C.). The early levels present occupation remains, while the later ones are of funerary nature. Lorestan Province, Western Iran. Tonoike 2013, 2014.

Huanbei
Huanbei preceded Yinxu as a city center of the Shang dynasty (13th century B.C.) in China. Its palace temple, which extended over 42 hectares, was destroyed 50 years after its construction. Stoltman *et al.* 2009; Stoltman 2016.

Jonathan Creek 15ML4C
Mississippian culture site occupied between A.D. 1200 and 1300. It was a large town-and-mound center, located in the west of the State of Kentucky in the United States. The site consisted of seven mounds, several plazas, houses and defensive walls. Schroeder 2009.

Kuntur Wasi
Ceremonial center of the Formative period in Peru (950–50 B.C.), located in the north central Andes at 2300 meters elevation, in the district of San Pablo, Department of Cajamarca. There are four main occupation levels, with important phases of architectural development, including platforms, temples and plazas. Several carved monoliths and the discovery of eight intact shaft tombs with many gold ornaments, ceramics and other precious items contributed to the fame of the site. Druc *et al.* 2013; Inokuchi 2010; Onuki *et al.* 1995; Onuki and Inokuchi 2011.

Mallard Bay Isle
Shell midden from the Coles Creek/Mississippian culture (A.D. 700–1200), Cameron County, south of the State of Louisiana, in the United States. Stamped ceramics with complex decoration generated an important study in which the question of the circulation of the wares or of the paddles to decorate them was addressed. See Saunders and Stoltman 1999 on this regard.

Mallas, see *Acopalca*

Mangallpa (Cuscuden)
Ceramic production center in the San Pablo District, Department of Cajamarca, Peru. Male activity. Utilitarian wares are produced with volcanic material and the paddle-and-anvil technique for manufacture and decoration. The Mangallpa potters are itinerant during part of the production season, which roughly spans from May till October, the dry months in the Andes. Druc 2011. Also http://vimeo.com/55308616.

Marcajirca
Defensive site of the Late Intermediate Period (A.D. 1000–1430) constructed at 3800 meters above sea level in the Callejon de Conchucos, Ancash Department, Peru. The site was occupied up to around 1640 at the end of the Colonial Period. There are ceremonial, public, funerary and residential sectors. Ibarra 2003.

Marcará/Musho/Cancharumí/Pariahuanca/Taricá
Production sites for utilitarian ware in the Callejón de Huaylas, Department of Ancash, Peru. Only a few potters are still producing utilitarian ware with the traditional paddle-and-anvil technique, on a seasonal or occasional basis. It is a male activity. The paste is produced by adding one of more sandy material(s) to the clay base. The major production center in the valley is Taricá, where many potters have switched to producing tourist ware with non-traditional techniques (wheel-thrown or modeled wares, painted decorations, varnish, electric kiln firing). Druc 1996; Druc and Gwyn 1998.

Mina Clavero
Ceramic production site in the Valley of Traslasierra, Province of Cordoba, Argentina. Few potters still follow the tradition, which uses thin coils to build the jars, cooking pots and other wares produced. See the video documentary about Atilio López filmed in 2001 (Druc 2012), and the documentary about his parents, Alcira and Jesús Tomás López produced in 1965 by Raymundo Gleyzer (Gleyzer and Montes de Gonzales 1965).

Pallka
Ceremonial center of the Andean Formative Period (first millennium B.C.), with residential and public areas, and a cemetery. Casma Valley, North-Central Coast, Department of Ancash, Peru. Druc 1998.

Pariahuanca, see *Marcará*

Phu Lang
Ceramic village 60 km Northeast of Hanoi, Vietnam, in the Red River Delta, specialized in stoneware production. Traditional techniques include the spinning wheel (now electric but initially a kick wheel) for elaborating cooking pots, jars, urns and vases (women activity), and slabs for making small coffins (men and women). Coiling is used jointly with the wheel for making jars, stacking one or two coils at a time and shaping the jar on the wheel. Red silty clay is used for elaborating all Phu Lang products, and river mud with bamboo (and other) leaf ash for the glaze yielding green to dark brown semi-vitrified coatings. Products are high fired with wood in large dragon-type kilns (long multichambered kilns). Druc 2014; Franchette and Stedman 2010; Stevenson and Guy 1997.

Puémape
Archaeological site of the Andean Formative Period on the Northern Coast of Peru, Department of La Libertad. Site affiliated to the Cupisnique culture, with ceremonial platforms, areas of domestic activity and cemeteries. Druc 2015; Elera 1998.

Pukará (or Pucará) de Tilcara
Archaeological site in the Humahuaca Quebrada, Norwestern Argentina. Located at 2525 meters above sea level on a promontory. The site dates from the 10th century A.D and was occupied up to the Inca occupation and the subsequent Hispano-Indigenious Period. It covers eight hectares and was densely constructed, with residential buildings, patios, plazas, corals, and cemeteries separated from the residential areas. The site was not defensive in character but the structures were grouped to form several large walled compounds with small plazas. Acevedo 2013; Tarragó 1992; Zaburlín 2009, 2010.

Qaleh Paswah 5 Irán
Archaeological site in the Valley of Qaleh Paswah, close to lake Urmia, Iran. Occupational levels date from the Dalma (Calcolithic - 6th millennium B.C.) and Islamic periods. Tonoike 2013, 2014.

San Marcos Acteopán
Ceramic production center in the State of Puebla, Mexico. Wares are produced with mold by pressing a tortilla-like clay slab unto it. The rim is done with a thin coil. Only one raw material is used from which two resources are extracted by decantation to separate the fine from the coarse sand fractions. The fine fraction and slurry is used as clay base, the coarse sand fraction is used as temper in varying proportions according to the ware to be produced. Druc 2000.

Sorkun
Ceramic production center in Central Anatolia, Turkey, between Eskisehir and Ankara. Cooking pots with a flat base and large opening are produced with the coiling technique, using ground micaceous schist or gypsum as temper. The women are in charge of all stages of the production. Druc 2008.

Taricá, see *Marcará*

Xiaomintun
Archaeological site where more than 1000 burials of the Shang dynasty were found (2nd millennium B.C.). The site is located in the modern city of Anyang, Henan Province, North Central China. Several individuals seem to have been linked to the production of bronze objects. Yinxu Xiaomintun Archaeological Team 2009.

Yacya, see *Acopalca*

Yinxu/Anyang
Last capital of the Shang dynasty occupied between 1200 and 1047 B.C. The site is located close to the modern city of Anyang, Henan Province, North Central China. Stoltman *et al.* 2009; Stoltman 2016.

GLOSSARY

botryoidal: globular or grape-like shape of a crystal, characteristic in particular of certain oxides and hydroxides, such as haematite, goethite, and malachite.

canchero: a small- to medium-size globular vessel form with a handle (like a Swiss fondue pot*).

ceramic: in this manual, a general term for any ware or object made of clay (mixed or not to a less plastic material), and transformed into a hard material through firing.

clay: refers to an ensemble of minerals (montmorillonite, illite, smectite, etc.) composed of phyllosilicates of aluminum of a size inferior to 2 μm (0.002 mm ISO scale). When mixed with water clay becomes plastic, and hardens when dried and fired. Clay deposits mined by traditional potters are rarely pure and can carry natural inclusions, organic and/or mineral, of fine to coarse sand size.

clay matrix: in ceramic analysis the term refers to the argillaceous matrix surrounding the other inclusions in the paste, organic or mineral.

clastic: fragment, made of fragments (of minerals or rocks).

cleavage: plane of structural weakness, usually parallel to the crystallographic planes of a mineral along which a mineral will break; preferential direction of rupture. E.g. slate is mined for roof tiles taking advantage of this property. The inter-cleavage angle of a mineral is a characteristic feature helping mineral identification.

decantation: process of separating the granulometric fractions or the finer from the coarser particles by letting them settle in water.

detritic (sediment): loose material consisting of fragments of rocks and minerals eroded from the Earth surface.

felsic: in geology, the term felsic is used to describe certain silica-rich, clear-colored minerals such as quartz and feldspars, and rocks formed with or rich in these minerals.

fresh paste or *fresh cut*: in this manual, 'fresh' means that the paste is examined from a recently broken-off ceramic fragment, allowing an observation of the cross-section free of deposit and surface contamination.

glaze: a nonporous, vitreous surface coating, fused onto the ceramic surface at high temperature (circa 1000-1200 °C or lower if fluxes are used), made from (semi)refractory clays, silica and alumina-rich raw materials, such as powdered glass, quartz, feldspars, and different oxides as fluxes and/or coloring agents.

grain: in this manual, a fragment or particle of mineral nature (as opposed to an inclusion that can be of any nature).

granulometry: grain size scale, used in paste analysis to classify mineral crystals, rock fragments and any other inclusions according to granulometric classes, often following the Udden-Wentworth scale (Folk 1965) into very fine, fine, medium, coarse and very coarse sand sizes.

grog: crushed ceramic fragments added as temper to the clay base.

inclusion: in ceramic paste analysis, the term refers to any non-argillaceous (and non-clay size) material in the paste.

leucocratic: in geology, a color index to describe rocks of clear color.

levigation: process of separating granulometric fractions or particles by using running water that will carry the lighter material further away than the heavier material. In ceramic production, potters may use an inclined surface or container and let water run through it to separate the fine from the coarse particles in their clay or temper materials.

lixiviation: leaching of the soluble constituents in a rock by water percolation.

mafic: a color index used in geology to describe certain minerals of dark-color (such as amphiboles, pyroxenes and micas). It also refers to minerals and rocks with a chemical composition rich in magnesium and iron.

matrix: in ceramic analysis, matrix refers to the clay base in which non-argillaceous and larger-than-clay-size particles or inclusions are found. In geology, the term matrix of a rock is the mass of fine grains in which larger crystals and lithoclasts are found.

meteorization: chemical alteration by rain water, alteration at the surface of the earth.

non-plastic material: inclusions/material that do/does not have the characteristics nor size of clay minerals; natural inclusions in the clay material or matrix of a size larger than clay minerals (superior to 2 μm – depending the granulometric scale).

paste: in ceramic analysis, paste refers to the mix of clay and non-plastic materials used in ceramic production. It can refer to the unfired body or fired product.

petrography: the study of the composition and texture of minerals and rocks in thin section with a polarizing microscope working with transmitted-light (shining through the sample from beneath the rotating stage). Ceramic petrography uses this technique to study the components and texture of a ceramic paste. Resolution of the petrographic microscope does not allow the identification of the clay minerals and inclusions of the size of clays (inferior to 0.02 mm).

phenocryst: in geology, a coarse crystal in a matrix of much smaller crystals, a characteristic feature of a porphyritic texture. Usually, phenocrysts are of the same minerals than found in the groundmass.

photomicrograph: photograph of a sample taken with a microscope.

plastic: malleable.

porosity: measure of the volume of pores or voids in a material.

prismatic: form (or habit) of a crystal which is elongated, has flat sides and presents faces that are parallel to the central axis.

sediment (clastic, detritic, etc.): a sediment results from the accumulation of crystals and rock fragments following weathering of preexisting rocks above the earth surface. It can be transported and deposited by the action of wind, water, or ice. Minerals and rock fragments in a sediment can be of different geological origin. A sediment can result from physical or chemical weathering processes or from the action of bioorganisms.

slip: a clay slurry or suspension of very fine materials used as surface coating on a ceramic. When applied too thick a slip will peel.

stoneware: a dense, nonporous, high-fired ceramic product elaborated with a clay rich in silt to fine sand-size quartz, feldspar and high-firing or refractory clay minerals, such as kaolin and illite able to withstand temperatures of circa 1200 °C.

temper: material added by the potter to the clay base, used to lower the plasticity of the clay, while giving structure and strength to the ceramic body. Used by a potter to modulate the plasticity of his/her paste.

texture (in ceramic analysis): aspect of a paste combining elements of distribution, size and percentage of inclusions, proportion of clay vs. natural inclusions or temper (if a conscious addition of a second material can be detected), and aspects of the clay matrix.

textures (in geology):

- *aphanitic*: index of grain size, indicative of a very fine crystalline texture, which cannot be distinguished with the naked eye.

- *aphyric*: in geology, the term refers to a fine-grained texture in igneous rocks, without phenocrysts.

- *blastic*: a texture of metamorphic origin indicative of recrystallization due to metamorphism.

- *cryptocrystalline*: very fine crystalline structure, a texture with microcrystals so fine as to be nearly indistinguishable with the microscope .

- *holocrystalline*: index of crystallinity used to designate rocks composed of more than 90 % crystals.

- *hypocrystalline:* texture term for volcanic rocks made up of crystals and glass.

- *phaneritic*: a grain size index, a reference to crystals that can be recognized with the naked eyes.

- *porphyritic*: bimodal texture with large crystals (phenocrysts) in a matrix of much smaller ones.

thin section: a very thin film of material (0.03 mm - 30 micron), in that case ceramic, which allows the light of a microscope to shine through. Used in ceramic petrography for the identification of minerals, rocks, and other inclusions in the paste larger than the clay particles.

twinning: in geology, crystal twinning is the symmetric grouping of identical crystals. It is a characteristic crystallographic feature of several minerals, and in particular plagioclases.

REFERENCES

Acevedo, V. J. 2013. Relevamiento y Análisis Arqueométrico de Materiales Cerámicos de Colecciones Situadas en Museo. *Actas del V Congreso Nacional de Arqueometría, Primer encuentro latinoamericano de tecnologías históricas y "metodologías científicas aplicadas al estudio de los bienes culturales"*, Universidad Nacional de Rosario - Universidad Tecnológica Nacional Rosario, octubre 23-25 2013, Santa Fe, Argentina.

Arnold, D. E. 1985. *Ceramic theory and cultural process*. Cambridge University Press, Cambridge.

Arnold, D. E. 2005. Linking society with the compositional analyses of pottery: A model from comparative ethnography. In *Pottery Manufacturing Processes: Reconstitution and Interpretation*, A. Livingstone Smith, D. Bosquet and R. Martineau (eds), pp. 15-21. BAR International Series 1349. Archaeopress, Oxford.

Cheel, R. J. 2005. *Introduction to clastic sedimentology*. Brock University, ON.

Cremonte, M. B. and L. Pereyra Domimgorena, L. 2013. *Atlas de pastas cerámicas arqueológicas. Petrografía de estilos alfareros del NOA*. Universidad Nacional de Jujuy, San Salvador de Jujuy, Argentina.

Cuadros, J., Afsin, B., Jadubansa, P., Ardakani, M., Ascaso, C. and J. Wierzchos. 2013. Pathways of volcanic glass alteration in laboratory experiments through inorganic and microbially-mediated processes. *Clay Minerals* 48: 423-445.

Druc, I. 1996. Entrevistas con ceramistas andinos: Inferencias para estudios de procedencias y caracterización cerámica. *Bulletin de l'Institut français d'études andines*, 25(1): 17-41.

Druc, I. 1998. *Ceramic production and distribution in the Chavín sphere of influence*. British Archaeological Reports, International Series 731, Oxford.

Druc, I. 2000. Ceramic production in San Marcos Acteopan, Puebla, Mexico. *Ancient Mesoamerica* 11: 77-89.

Druc, I. 2005. *Producción cerámica y etnoarqueología en Conchucos, Ancash, Perú*. Instituto Runa, Lima.

Druc, I. 2008. *Women potters of Sorkun, Turkey*. Documental, 12 min. Poiesis Creations. www.vimeo.com. http://vimeo.com/35526032.

Druc, I. 2011. Tradiciones alfareras del valle de Cajamarca y cuenca alta del Jequetepeque, Perú. *Bulletin de l'Institut français d'études andines* 40(2):

307-331.

Druc, I. 2012. *Atilio López, alfarero tradicional de la sierra argentina.* Video documentary, 18 min. Poiesis Creations. www.vimeo.com. https://vimeo.com/97191953.

Druc, I. 2014. Phu Lang stoneware production, Northern Vietnam: Raw materials and fired products. Text of the video documentary: *Phu Lang: A ceramic village in Northern Vietnam.* (www.vimeo.com, https://vimeo.com/134255181). Poiesis Creations, WI.

Druc, I. 2015. Charophytes in my plate: Ceramic production in Puemape, North Coast of Peru. In *Ceramic Analysis in the Andes: Proceedings of the session on Andean Ceramic Characterization, Society for American Archaeology Annual Meeting 2014, Austin Texas,* I. Druc (ed.), pp. 37-56. Deep University Press, WI.

Druc, I., Burger, R. L., Zamojska, R., and P. Magny. 2001. Ancón and Garagay Ceramic Production at the Time of Chavín de Huántar. *Journal of Archaeological Science* 28(1): 29-43.

Druc, I. and H. Gwyn. 1998. From clay to pots: A petrographic analysis of ceramic production in the Callejón de Huaylas, North-Central Andes, Peru. *Journal of Archaeological Science 25*(7)*:* 707-718.

Druc, I., Inokuchi, K. and Z. Shen. 2013. Análisis de arcillas y material comparativo para Kuntur Wasi, Cajamarca, Perú por medio de difracción de rayos X y petrografía. *Arqueología y Sociedad* 26: 91-110.

Elera A. C. 1998. The Puémape site and the Cupisnique culture: A case study on the origins and development of complex society in the Central Andes, Peru. Dissertation thesis. Department of Archaeology, University of Calgary, Alberta, Canada.

Folk, R.L. 1951. A comparison chart for visual percentage estimation. *Journal of sedimentary research* 21 (1): 32-33.

Folk, R. L. 1965. *Petrology of sedimentary rocks.* The University of Texas, Austin.

Franchette, S., and N. Stedman. 2010. *A la découverte des villages de métier au Vietnam. Dix itineraires autour de Ha Noi.* IRD Institut de recherche pour le dévelopement, Marseille, France.

Gleyzer, R. and A. Montes de Gonzales. 1965. *Ceramiqueros de Traslasierra.* Video documentary, 19 min. Escuela de Artes, Universidad Nacional de Córdoba, Argentina.

Goren, Y. 2014. The operation of a portable petrographic thin section laboratory for field studies. *New York Microscopical Society Newsletter*, September on-line issue.

Gosselain, O. 1992. Technology and style: Potters and pottery among the Bafia of Cameroon. *Man* 27(3): 559-586.

Gosselain, O. 2000. Materializing identities: An African perspective. *Journal of Archaeological Method and Theory* 7(3): 187-217.

Gosselain, O. 2008. Thoughts and adjustments in the potter's backyard. In *Breaking the mould: Challenging the past through pottery*, I. Berg (ed.), pp. 67-79. BAR International Series 1861. Archaeopress, Oxford.

Ibarra Asencios, B. 2003. Arqueología del valle del Puchca. In *Arqueología de la Sierra de Ancash,* B. Ibarra Asencios (ed.), pp. 252-330. Instituto Cultural Runa, Lima.

Inokuchi, K. 2010. La arquitectura de Kuntur Wasi: secuencia constructiva y cronología de un centro ceremonial del Periodo Formativo. *Boletín de Arqueología, Pontificia Universidad Católica del Perú* (PUCP) 12: 219-248.

Kingery, W. D. 1982. Plausible inferences from ceramic artifacts. In *Archaeological Ceramics*, J. S. Olin and A. D. Franklin (eds), pp. 37-45. Smithsonian Institution Press, Washington, D.C.

Kirov, B. L. and N. N. Truc. 2012. A study on the relationship between geotechnical properties and clay mineral composition of Hanoi soft soils in saline media. *International Journal of Civil Engineering* 10(2): 87-92.

Livingood, P. C. and A. S. Cordell. 2009. Point/Counter Point: the Accuracy and Feasibility of Digital Image Techniques in the Analysis of Ceramic Thin Sections. *Journal of Archaeological Science* 36: 867-872.

Livingood, P. C. and A. S. Cordell. 2016. Point/Counter Point II: The Accuracy and Feasibility of Digital Image Techniques in the Analysis of Pottery Tempers Using Sherd Edges. In *Integrative Approaches in Ceramic Petrography*, M. Ownby, I. Druc and M. Masucci (eds) (pp. tba). University of Utah Press, Salt Lake City.

MacKenzie, W. S., Donaldson, C. H. and C. Guilford. 1991. *Atlas of igneous rocks and their textures.* Longman Scientific & Technical. John Wiley & Sons, New York.

Marsh, L. and Druc., I. 2015. *Sampling paste for thin section: An Andean case study of the initial steps of petrographic research.* In *Ceramic Analysis in the Andes.* I. Druc (ed.), pp. 157-170. Deep University Press, WI.

Matson, F. R. 1970 [1963]. Some aspects of ceramic technology. In *Sciences in Archaeology*, D. Brothwell and E. Higgs (eds), pp. 592-601. Thames and Hudson, London.

Matthew, A. J., Woods A. J., and C. Oliver, 1997. Spots before the eyes: new comparison charts for visual percentage estimation in archaeological material. In *Recent Developments in Ceramic Petrology*, A. P. Middleton and I. C. Freestone (eds), pp. 211-264. British Museum Occasional Paper No. 81. British Museum Press, London.

Middleton, A. P., Freestone, I.C., and M. N. Leese, 1985. Textural analysis of ceramic thin sections: Evaluations of gain sampling procedures. *Archaeometry* 27(1): 64-74.

Middleton, A. P., Leese, M. N., and M. R. Cowell. 1991. Computer-assisted approaches to the grouping of ceramic fabrics. In *Recent Developments in Ceramic Petrology*, A. P. Middleton and I. C. Freestone (eds), pp. 265-267. British Museum Occasional Paper No. 81. British Museum Press, London.

Onuki, Y. and K. Inokuchi. 2011. *Gemelos Pristinos: el tesoro del templo de Kuntur Wasi*. Fondo editorial Congreso del Perú, Lima.

Onuki, Y., Kato, Y. and K. Inokuchi. 1995. La primera parte: Las excavaciones en Kuntur Wasi, la primera etapa, 1988-1990. In *Kuntur Wasi y Cerro Blanco*, Y. Onuki (ed.), pp. 1-126. Hokusen-Sha, Tokyo.

Perkins, D. 2002. *Mineralogy*. Prentice Hall, New Jersey.

Reedy, C. L. 2006. Review of Digital Image Analysis of Petrographic Thin Sections in Conservation Research. *Journal of the American Institute for Conservation* 45(2): 127-146.

Rice, P. M. 1987. *Pottery analysis: A source book*. University of Chicago Press, Chicago.

Roduit, N. 2002-2008. JMicroVision v.1.2.7. www.jmicrovision.com.

Rosas La Noire, H. 2007. *La secuencia cultural del período formativo de Ancón*. 1. ed. Serie Tesis. Avqi Ediciones, Perú.

Rye, O. 1981. *Pottery Technology. Principles and Reconstruction*. Taraxacum, Washington.

Saunders, R. and J. Stoltman. 1999. A Multidimensional consideration of complicated stamped pottery production in Southern Louisiana. *Southeastern Archaeology* 18(1): 1-23.

Schroeder, S. 2009. Viewing Jonathan Creek through ceramics and radiocarbon dates: Regional prominence in the thirteenth century. In *TVA Archaeology. Seventy-five Years of Prehistoric Site Research*, E. E. Pritchard and T. M. Ahlman (eds), pp. 145-180. The University of Tennessee Press, Knoxville.

Shepard, A. O. 1942. *Rio Grande glaze-paint ware: A study illustrating the place of ceramic technological analysis in archaeological research*. Publication 526, Contributions to Anthropology 39. Carnegie Institution of Washington, Washington, D.C.

Shepard, A. O. 1964. Temper identification: "Technological sherd-splitting" or an unanswered challenge. *American Antiquity* 29(4): 518-520.

Shepard, A. O. 1965. Rio Grande glaze-paint pottery: A test of petrographic analysis. In *Ceramics and Man*, F. R Matson (ed.), pp. 62-87. Aldine, Chicago.

Shepard, A. O. 1968 [1956]. *Ceramics for the archaeologist*. Carnegie Institution of Washington, Washington, D.C.

Stevenson J. and J. Guy. 1997. *Vietnamese ceramics. A separate tradition*. Art Media Resources with Avery Press, Chicago.

Stoltman, J. 1989. A quantitative approach to the petrographic analysis of ceramic thin sections. *American Antiquity* 54: 147-160.

Stoltman, J. 1999. The Chaco-Chuska Connection: In Defense of Anna Shepard. In *Pottery and People. A Dynamic Interaction*, J. M. Skibo and G. M. Feinman (eds), pp. 9-24. Foundation for Archaeological Inquiry, Utah Press, Utah.

Stoltman, J. 2016. The Use of loess in pottery manufacture: A comparative analysis of pottery from Yinxu in North China and LBK sites in Belgium. In *Integrative Approaches in Ceramic Petrography*, M. Ownby, I. Druc and M. Masucci (eds), (pp. tba). University of Utah Press, Salt Lake City.

Stoltman, J., Zhichun J., Jigen T., and G. (Rip) Rapp. 2009. Ceramic Production in Shang Societies of Anyang. *Asian Perspectives* 48(1): 182-203.

Strienstra, P. 1986. Systematic macroscopic description of the texture and composition of ancient pottery. Some basic methods. University of Leiden, Department of pottery technology. *Newsletter* 4: 29-48.

Tarragó, M. 1992. Áreas de actividad y formación del sitio de Tilcara. *Cuadernos* 3: 64-74.

Tonoike, Y. 2012 (published 2014). Using Petrographic Analysis to Study the 6th Millennium B.C. Dalma Ceramics from Northwestern and Central Zagros.

Iranian Journal of Archaeological Studies 2(2): 65-82.
http: //ijas.usb.ac.ir/?_action=article&au=7285&_au=Yukiko++Tonoike

Tonoike, Y. 2013. Beyond Style: Petrographic Analysis of Dalma Ceramics from Two Regions in Iran. In *Interpreting the Late Neolithic in Upper Mesopotamia* (Publications on Archaeology of the Leiden Museum of Archaeology), P.M.M.G. Akkermans, O. Nieuwenhuys, and R. Bernbeck (eds), pp. 397-406. Brepols Publisher, Belgium.

Velde, B., and I. Druc. 1999. *Archaeological Ceramic Materials. Origin and Utilization.* Springer-Verlag, Berlin & New York.

Weigand, P. C., Harbottle, G., and E. V. Sayre. 1977. Turquoise sources and source analysis: Mesoamerica and the Southwestern U.S.A. In *Exchange Systems in Prehistory,* T. K. Earle and J. E. Ericson (eds), pp. 15-34. Academic Press, New York.

Willey, G. R. and J. M. Corbett. 1954. *Early Ancón and early Supe culture, Chavín horizon sites of the Central Peruvian coast.* Columbia Studies in Archeology and Ethnology, Columbia University Press, New York.

Winter, J. D. 2010 (2nd ed). *An introduction to igneous and metamorphic petrology.* Prentice Hall, New York.

Whitbread, I. K. 1986. The characterisation of argillaceous inclusions in ceramic thin sections. *Archaeometry* 28(1): 79-88.

Whitbread, I. K. 1991. Image and data processing in ceramic petrology. In *Recent developments in ceramic petrology,* A. Middleton and I. Freestone (eds), pp. 369-86. British Museum Occasional Paper No. 81, London.

Yinxu Xiaomintun Archaeological Team. 2009. 2003-2004 Excavation of Shang tombs at Xiaomintun in Anyang City, Henan. *Chinese Archaeology* 9(1): 90-98.

Zaburlín, M. A. 2009. Historia de ocupación del Pucará de Tilcara (Jujuy, Argentina). *Intersecciones en Antropología* 10: 89-103.

Zaburlín, M. A. 2010. Las reconstrucciones arqueológicas analizadas como discursos sobre el pasado. *Revista de arquitectura TRAMA* no 103, digital edition, September, article no 376.

PASTAS CERÁMICAS EN LUPA DIGITAL:
COMPONENTES, TEXTURA Y TECNOLOGÍA
Isabelle Druc y Lisenia Chavez

Este manuel es el primer texto basado en el uso de las nuevas lupas digitales portátiles para análisis de cerámicas arqueológicas en trabajos de campo y laboratorio. Con más de 90 ilustraciones en color, se presenta como un Atlas de geología, con identificación de los minerales y fragmentos líticos comunes en pastas cerámicas. Es una ayuda metodológica para los arqueoólogos, para la clasificación de los fragmentos de cerámica, los estudios estilísticos y la selección de las piezas más representativas para estudios químicos y petrográficos posteriores. Este manual permite identificar muchas de las inclusiones que se ven con una lupa, así como varios elementos de tecnología y manufactura cerámica. Se propone un protocolo de análisis y nociones de granulometría y geología que permiten la constitución de grupos de composición y textura similares para alcanzar un mayor nivel interpretativo de los datos cerámicos en arqueología.

.....

"El libro de Druc y Chavez es una contribución muy grande para los arqueólogos, porque nos lleva a compartir la metodología y la terminología para el análisis de las pastas cerámicas, mostrandonos nuevas posibilidades de estudio de las cerámicas".

—*Kinya Inokuchi*
Universidad de Saitama, Japon

"Manual indispensable para todo arqueólogo interesado en brindar sustento sólido y objetivo a la discusión de la organización del sistema de producción y distribución de cerámica mediante el estudio de la cadena operativa. Contribución importante en pos de uniformizar métodos de clasificación de material cerámico y la selección de muestras para análisis arqueométricos avanzados. Aportará sin duda a la mayor solidez de la discusión sobre las identidades de productores y usuarios, estas mismas que se expresan en las decisiones tecnológicas y estilísticas que fueron tomadas por el alfarero".

—*Krzysztof Makowski*
Pontificia Universidad Católica del Perú

ISBN: 978-1939755049

ALSO AT DEEP UNIVERSITY PRESS:

Isabelle C. Druc Ed. (2015)

Ceramic Analysis in the Andes

CONTENT

AVAILABLE HERE: http://www.deepuniversitypress.org/ceramic.html

Also by...

Druc, I. 2005. *Producción cerámica y etnoarqueología en Conchucos, Ancash, Peru.* Instituto Runa, Lima.

Druc, I. (Ed.) 2001. *Archaeology and Clays.* British Archaeological Reports S942. Adrian Books, Oxford.

Ownby, M., Druc, I. and M. Masucci (Eds). 2016. *Integrative Approaches in Ceramic Petrography.* University of Utah Press, Utah.

Velde, B., and I. Druc. 1999. *Archaeological Ceramic Materials. Origin and Utilization.* Springer-Verlag, Berlin & New York.

YOU MIGHT WANT TO READ:

SIGNS AND SYMBOLS IN EDUCATION

François Victor Tochon, Ph.D.

University of Wisconsin-Madison, USA

In this monograph on Educational Semiotics, François Tochon (along with a number of research colleagues) has produced a work that is truly groundbreaking on a number of fronts. First of all, in his concise but brilliant introductory comments, Tochon clearly debunks the potential notion that semiotics might provide yet another methodological tool in the toolkit of educational researchers. Drawing skillfully on the work of Peirce, Deely, Sebeok, Merrell, and others, Tochon shows us just how fundamentally different semiotic research can be when compared to the modes and techniques that have dominated educational research for many decades. That is, he points out how semiotic methods can provide the capability for both students and researchers to look at this basic and fundamental human process in inescapably transformational ways, by acknowledging and accepting that the path to knowledge is, in his words "through the fixation of belief."

But he does not stop there – instead, in four brilliantly conceived studies, he shows us how semiotic concepts in general, and semiotic mapping in particular, can allow both student teachers and researchers alike insights in these students' development of insights and concepts into the very heart of the teaching and learning process. By tackling both theoretical and practical research considerations, Tochon has provided the rest of us the beginnings of a blueprint that, if adopted, can push educational research out of (in the words of Deely) its entrenchment in the Age of Ideas into the new and exciting frontiers of the Age of Signs.

Gary Shank
Duquesne University

SEE REVIEWS HERE: http://www.deepuniversitypress.org/signs.html

Out of Havana:
Memoirs of Ordinary Life in Cuba

Dr. Araceli Alonso
University of Wisconsin-Madison

Out of Havana provides an uncommon ordinary woman's insight into the last half century of Cuba's tumultuous recent history. More powerfully than an academic study or historical account, it allows us intimately to grasp the enthusiasm, commitment and sense of promise that defined many average Cubans' experience of the 1959 Revolution and the first triumphant decades of the Castro regime. As the story shifts into the final decades of the last century (the 1980s Mariel Boatlift, the so-called "special period in time of peace" [from 1991 to the end of the decade], and the 1994 Balseros or Rafters Crisis), it starts gradually to reveal, with understated yet relentless eloquence, an ultimately insuperable rift between the high-flown official rhetoric of uncompromising struggle and revolutionary sacrifice and the harsh conditions and cruelly absurd situations that the protagonist, along with the majority of Cubans, begin routinely to live out. It is a rare and important document, a unique personal chronicle of an everyday Cuban reality that most Americans continue to know only fragmentarily.

Dr. Araceli Alonso is a 2013 United Nations Award Winner for her activism on women's health and women right. Associate Faculty at the University of Wisconsin-Madison in the Department of Gender and Women's Studies and in the School of Medicine and Public Health, she is the Founder and Director of the award-winning non-profit organization Health by Motorbike.

http://deepuniversitypress.org/havana.html

Science Teachers Who Draw:
The *Red* Is Always There

Dr. Merrie Koester
Project Draw for Science
Center for Science Education
University of South Carolina

This book documents the ways in which science teacher researchers used drawing to construct semiotic spaces inside which students acquired significant aesthetic capital and agency. Many previously failing students brokered this new capital into improved academic achievement and a sense of felt freedom.

Science Teachers Who Draw: The Red is Always There is a book which asks, "What happens when science teachers adopt an *aesthetic* approach to inquiry, using drawing to communicate deep understanding?" This narrative inquiry was driven by quantitative studies which reveal a robust positive correlation between students' test scores in reading and science, beginning at the middle school level. When the data are disaggregated, there exists a vast achievement gap for low income and English language learners. Science teachers are faced with a semiotic nightmare.   Often possessing inadequate pedagogical content knowledge themselves, science teachers must somehow symbolically *communicate* often highly abstract knowledge in ways that can be not only be decoded by their students' but later used to construct deeper, more differentiated knowledge, which can be applied to make sense of and adapt successfully to life on Planet Earth.

An invaluable resource for teachers, teacher educators, and qualitative researchers.

http://www.deepuniversity.net/koester.html

http://www.deepuniversitypress.org/red.html

Formación Y Desarrollo De Profesionales De La Educación: Un enfoque profundo

Manuel Fernández Cruz

Universidad de Granada

El libro contiene herramientas prácticas para la intervención formativa. Se plantea la formación desde la perspectiva del desarrollo profesional y se adopta un enfoque profundo novedoso que integra los ámbitos racionales, emocionales y vivenciales que requiere el aprendizaje y la actualización permanente en las profesiones educativas: la docencia, la pedagogía, la psicología, o la formación de formadores.

- El estudiante universitario va a contar con un manual teórico-práctico de referencia para dominar el ámbito de la formación y el desarrollo profesional.

- El educador va a encontrar en el texto herramientas provechosas para encarar su propio proceso de desarrollo y perfeccionamiento.

- El formador de formadores va a disponer de referencias teóricas y actividades prácticas que facilitarán su intervención.

Dr. Manuel Fernández Cruz, Licenciado en Pedagogía y Doctor en Ciencias de la Educación, es profesor de la Universidad de Granada desde 1992. Actualmente es Director del Departamento de Didáctica y Organización Escolar y Coordinador General del Consorcio Internacional MUNDUSFOR (Formación de Profesionales de la Formación).

http://www.deepuniversitypress.org/formacion.html

TRANSFER OF LEARNING
AND THE CULTURAL MATRIX

Culture, Beliefs and Learning
in Thailand Higher Education

Dr. Jonathan H. Green

University of Southern Queensland

The field of quality teaching and learning is a complex and dynamic one. Jonathan Green's book on the transfer of learning makes an original contribution to this field in that it adds value to the discourse on influences and forces impacting on quality student learning. Learning is not a one-directed process, characterised by teacher-centeredness, but one where students are at the centre. Understanding how students perceive and experience their own learning is a key to unlocking their potential. This is a long-overdue publication.

—*Professor Arend E. Carl, Vice-Dean: Teaching, Stellenbosch University, South Africa*

Through this research, Jonathan Green has contributed to the body of knowledge about transfer of learning. His rigorous research investigates transfer in the context of learners' personal epistemology and culture, yielding a culturally relative understanding of transfer that is highly relevant in today's increasingly diverse classrooms. The findings, which have implications for educators in a wide range of educational contexts, will be of particular interest to those who teach in internationalized and multicultural institutions.

—*Alexander Nanni, Director, Preparation Center for Languages and Mathematics at Mahidol University International College, Thailand*

Original thought provoking, high quality research that extends our knowledge of transfer of learning in relation to multicultural tertiary students in international education settings. Deep insights are gained, through use of the researcher's Measure of Academic Literacy (MALT), a new tool that explored issues of context and cultural values and beliefs, and metacognitive knowledge in transfer of learning.

—*Associate Professor Shirley O'Neill, Applied Linguistics Discipline Coordinator, School of Linguistics, Adult and Specialist Education, University of Southern Queensland*

http://www.deepuniversitypress.org/transfer.html

FROM TRANSNATIONAL LANGUAGE POLICY TRANSFER TO LOCAL APPROPRIATION

The case of the National Bilingual Program in Medellín, Colombia

Dr. Jaime Usma Wilches

University of Antioquia

Embracing a critical and sociocultural perspective for the study of policy, this vertical case study investigates foreign language education policies being adopted by the national government in Colombia, and how they are reinterpreted and appropriated by local official and public school teachers in the city of Medellín.

ISBN 978-1-939755-20-9 (hb)

Maria Alfredo Moreira, University of Minho, Portugal:

Drawing on the example of Medellín, Colombia, Jaime Usma's book does a magnificent work at dismantling one of the most pervasive grand narratives in globalized transnational foreign language policies: proficiency in English as one of the strongest pillars of a vibrant modern knowledge society, associated with higher economic gains for all. The author cogently demonstrates how apparently neutral and technically sound transnational and national policymaking fails to properly address structural inequality and social and economic injustice, while being creatively reenacted by local schools and actors that appropriate them according to their own goals, needs, and desires towards a more just and humane society.

Doris Correa, Associate Professor, Universidad de Antioquia, Medellin, Colombia:

Usma's book is much more than a nice rendition of how transnational language policies are being appropriated by government officials and other educational actors in Colombia. It is an enjoyable journey into Colombia's most recent political, socio-economic and educational reforms, and a compelling critical analysis of how those reforms are influencing school and classroom practices in Medellin.

Anikó Hatoss, University of New South Wales, Sydney, Australia:

This volume represents much needed scholarship in exploring the tensions between official language policies and their practical implementation on the ground. While these tensions have long been the interest of language policy studies, it is rare that scholars explore them through such rigorous and multilayered empirical research capturing the global and local. The comparative and critical lens applied here makes this volume an outstanding contribution to the field and provides invaluable insights for researchers, policy makers, curriculum planners and language teaching practitioners about language education in Columbia with lessons to be learnt far beyond.

Book Series on Deep Research Methodologies

Research methodologies need to be reconceptualized in two ways: first, as the expression of dynamic interpretive prototypes that can be activated through deep forms of inquiry that go beyond the surface level at which meanings are essentialized and reified. Second, integrating emergent technologies, structure and agency to meet deeper, humane aims. The dynamism of human interpretation is meaning-producing through multiple connected intentions among disciplinary domains.

By tackling both theoretical and practical research considerations, this book series provides the readers a blueprint that can push research into the new and exciting frontiers of the Age of Signs (in the words of Gary Shank). Taking into account adaptive and complex situations is the prime focus of such a hermeneutic inquiry.

The intent of this book series is to propose instruments to analyze beyond the surface of the matter in favor of value-loaded investigations chosen in order to revolutionize the current state of affairs, in increasing our sense of responsibility for our actions as humans vis-à-vis our fellow humans and our home planet.

For more, see here:

http://www.deepuniversitypress.org/deep-research-methodologies.html

Language Education Policy Book Series

Language Education Policy (LEP) is the process through which the ideals, goals, and contents of a language policy can be realized in education practices. Language policies express ideological processes. Their analysis reveals the perceptions of realities proper to certain sociocultural contexts. LEPs further their ideologies by defining and disseminating the values of policymakers. Because Language Education Policies are related to status, ideology, and vision of what society should be and traditions of thoughts, such issues are complex, quickly evolving, submitted to trends and political views, and they need to be studied calmly. The way to approach them is to get comparative information on what has been done in many settings, which are working or not, which are their flaws and merits, and try to grasp the contextual variables that might apply in specific locations, without generalizing too fast.

Policy discourses and curricula reveal the ideological framing of the constructs that they encode and create, project, enact, and enforce aspects such as language status, power and rights through projective texts generated to forward and describe the contexts of their enactments. Policy documents are therefore socially transformative through their evaluative function that frames and guides action in order to achieve language reforms. While temperance and reflection are required to address such complex issues, because moving to fast may create trouble, nonetheless the absence of action in this domain may lead to systemic intolerance, injustice, inequity, mass discrimination and even, genocidal crimes.

http://www.languageeducationpolicy.org

http://www.deepuniversitypress.or g/language-education-policy.html

Guide to Authors

What our Publishing Team can offer:

➢ An international editorial team, in more than 20 universities around the world.

➢ Dedicated and experienced topic editors who will review and provide feedback on your initial proposal.

➢ A specific format that will speed up the production of your book and its publication.

➢ Higher royalties than most publishers and a discount on batch orders.

➢ Global distribution and marketing in the U.S., UK, Europe, Australia, Brazil, China, Japan, South Korea and other countries.

➢ Fast recognition of your work in your area of specialization.

➢ Quality design and affordable sales pricing. Using the latest technology, our books are produced efficiently, quickly and attractively.

➢ A global marketing plan, including electronic and web marketing and review mailing.

➢ Book Series: Deep Education; Deep Language Learning; Signs & Symbols in Education; Language Education Policy; Deep Professional Development; Deep Activism.

> ➢ http://www.deepuniversitypress.org/guide.html

> ➢ **Contact : publisher@deepuniversity.net**

Deep University Online !

For updates and more resources

Visit the Deep University Website:

www.deepuniversity.net

www.deepunversitypress.org

Contact: publisher@deepuniversity.net

❖ Online Certificate and Courses on Deep Education:
http://www.deepuniversity.net/graduatecourses.html

Correspondence for this manual:

Isabelle C. Druc, Department of Anthropology, 5240 Social
Science, 1180 Observatory Dr., University of Wisconsin—
Madison, Madison, Wisconsin 53706 USA.

E-mail: icdruc@wisc.edu

DEEP UNIVERSITY PRESS
SCIENTIFIC BOARD MEMBERS

Biographies

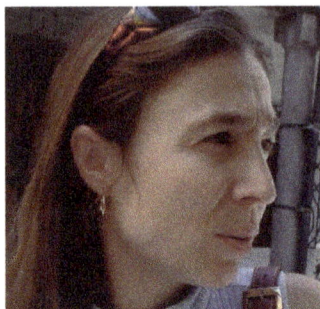

Isabelle C. Druc did her Ph.D. in Archaeology at the University of Montreal (Quebec, Canada), after finishing her initial studies in Switzerland. She is specialized in ceramic studies, Andean archaeology, and ethnoarchaeological research. She did her post-doctoral studies at Yale University in the United States, and has been a visiting scholar at the CNRS in France and at the Smithsonian in Washington D.C. She has received two excellence awards from the University of Montreal in Canada and won the 1989 Plantamour-Prévost science prize in Switzerland for her master at the University of Geneva. She has been at the University of Wisconsin-Madison since 2000, holding positions of lecturer and honorary fellow in the Department of Anthropology, and associate researcher in the Wisconsin Center for Education Research (WCER). She has been involved in many archaeological projects and ethnographic studies in South America, the USA, Europe, the Middle East and Southeast Asia, has published thirty articles and eight books as author, co-author, or editor, and has produced some 200 film documentaries and video interviews related to culture, ceramics, traditional arts, handicrafts, and language.

Bruce Velde is currently *Directeur de Recherche Emérite* with the *Centre National de la Recherche Scientifique* (CNRS) of France where he has worked since 1965 after a PhD from Montana State University and a post doctoral period with the Carnegie Geophysical Laboratory in Washington DC. He has published 208 articles as well as seven books as author or with co-workers and directed 21 theses. Fields of interest for his work have focused upon silicate mineral stabilities and properties under various geological conditions with special attention to clay minerals. However only recently has he focused his attention on clays at the surface.

Lisenia Chavez is a geologist specialized in Andean geology, with a degree in geology from the University of San Marcos in Lima, Peru (Universidad Nacional Mayor de San Marcos (UNMSM). She has worked at the Geology Institute in Lima (Instituto Geológico Minero y Metalúrgico – INGEMMET and in the area of geochemistry, petrology and petrography of igneous, metamorphic and sedimentary rocks. She also worked in geostatistic and in geology for the mining industry.

9 781939 755070